ALL THINGS WISE AND WONDERFUL, THE LORD GOD MADE THEM ALL

A Challenge to the Theory of Evolution for Latter-day Saints

D. JOHN REDMAN

COPYRIGHT AND DISCLAIMER

Cover design by D. John and Pam Redman
Book design and production by D. John Redman
Editing by Pam Redman and Kristen Bradshaw

All Things Wise and Wonderful, The Lord God Made Them All by D. John Redman

Published by HieToKolob Publishing

ISBN: 978-1-7357057-1-2 (pbk.)

ALL THINGS BRIGHT AND BEAUTIFUL[1]

REFRAIN

All things bright and beautiful,
All creatures great and small,
All things wise and wonderful,
The Lord God made them all.

VERSES

Each little flow'r that opens,
Each little bird that sings,
He made their glowing colors,
He made their tiny wings:

The purple-headed mountain,
The river running by,
The sunset and the morning,
That brighten up the sky:

The cold wind in the winter,
The pleasant summer sun,
The ripe fruits in the garden,
He made them ev'ry one:

The tall trees in the greenwood,
The meadows where we play,
The rushes by the water,
We gather ev'ry day:

He gave us eyes to see them
And lips that we might tell
How great is God Almighty
Who has made all things well:

---

[1] *All Things Bright and Beautiful*, Children's Songbook, p. 231, The Church of Jesus Christ of Latter-day Saints, words and music by Cecil Frances Alexander, 1818-1895; Hymns for Little Children, 1848, public domain.

TABLE OF CONTENTS

LIST OF FIGURES

LIST OF TABLES

PREFACE

Many members of The Church of Jesus Christ of Latter-day Saints have nagging questions concerning Darwinian Evolution. There are numerous articles and books on the subject, many diametrically opposed to each other. It is difficult for Saints to know who to trust. Unfortunately, once one gets past the principles of evolution taught in introductory high school or college biology classes, things get complicated quickly. One must transition from a few simple rules to the details of biochemistry, anatomy, physiology, genetics, cell biology, thermodynamics, mathematics, geology, and paleontology. Many of the books about evolution approach evolution from a single topic point-of-view such as DNA or the Cambrian Explosion. Consequently, one would need to read many full-length books. **This book is an attempt to assemble in one place, in a minimally necessary but sufficient scope, pertinent information that refutes naturalistic evolution and promotes the involvement of the divine.**

There are two paradigms of thought regarding evolution: (1) the chance-based, random, non-purposeful, naturalistic origin and evolution of life-forms paradigm; or (2) the creationist or intelligent design paradigm, the view that the origin, shape, and function of all life-forms are a result of a purposeful, planned, non-random, and designed process. These paradigms extend to individuals' worldviews (their beliefs and their attitudes) which, in turn, shape the way they think and act. This book will contrast the two paradigms throughout.

This book also discusses the often derogatory, and sometimes reviling attitude of many of the major evolutionary evangelists towards Christianity and God. Included is a significant collection of atheistic statements from these pro-evolutionists.

As the book progresses, the doctrine of The Church of Jesus Christ of Latter-day Saints arises to place the origin and development of life into a doctrinal context. If desired, the later chapters can be skipped without harming the preceding arguments. However, all Christians should find the overall discussion worthwhile. Periodically, Charles Darwin is compared to Isaac Newton and to the Prophet Joseph Smith as a means of contrasting their beliefs, teachings, and their impacts on the world. A watch or clockworks and biological machines are used as archetypes of purposeful design.

Emphasis is used throughout the text to emphasize key ideas. ***Special emphasis*** is given to discussions about events happening at random or by chance versus by Intelligent Design. When a reference to a significant author is first used, information about that author is included in the footnotes in addition to the reference.

I have always appreciated having references associated with the text immediately accessible without having to go to the back of the chapter or the book. Consequently, all references appear inline as footnotes on the same page that they are referenced.

I alone am responsible for the contents of this work. My views are not endorsed by The Church of Jesus Christ of Latter-day Saints. That said, I have made every effort to represent Church doctrine as accurately as possible. Thanks are due to many for reading the manuscript and providing suggestions. The foremost is my soulmate and eternal companion, Pam. Thanks also to my daughter, Rachel, Program Director and Assistant Professor for Medical Lab Sciences at a major university, and my son, Joseph, MD (hepatology)/Ph.D. (biochemistry), and a friend, Kristen Bradshaw, an engineer who translates technical documents into Japanese.

As I was finishing this book, I visited with a former colleague from Hewlett-Packard, Co. I told him I had just written a 250-page book for Latter-day Saints challenging Darwinian evolution. He said, “It takes 200 pages to say, ‘God did it!’?” My answer was “yes if I wanted to explain why evolution could not do it.” So, let us begin.

1. INTRODUCTION, THESIS & OUTLINE

Ps. 139:14 "… I am fearfully[2] and wonderfully made …"

In the spring of 1830, the Prophet Joseph Smith published the Book of Mormon and established what was to become known as The Church of Jesus Christ of Latter-day Saints. He claimed to have seen God the Father and His son, Jesus Christ, and a host of angels in preparation for restoring The original Church of Jesus Christ in the Latter-days. The Prophet taught that God the Father was the literal father of men's spirit bodies, that Adam and Eve were created in the image of God the Father and God the Mother, and that man's creation was purposeful – to become like God (Moses 1:39). Because of his claim of having seen visions and having received revelations, he was persecuted all the days of his life until he was killed by a mob at the age of thirty-nine years. The Prophet was secretly buried in a hidden grave to prevent his body from being desecrated by the enemies of righteousness. The next year, Charles Darwin boarded the HMS Beagle on his fateful scientific journey. Darwin had a burning desire to contribute to the newly developed natural philosophy wherein the laws of nature were to be discovered by observation and inductive reasoning. His observations and reasoning led to his theory that men are descended from lower forms of life through an exceedingly long sequence of chance, random, unguided, blind, and purposeless variations, subsequently acted upon by natural selection. Man, in other words, was a product of chance, and the idea of a God, particularly a loving Father-in-Heaven was unnecessary and injurious to his theory. Darwin became an instant celebrity and was held in high esteem by his fellowmen his entire life. He died at the age of seventy-three and his burial in Westminster Abbey was attended by thousands of family, friends, scientists, philosophers, and dignitaries. Darwin was buried in a place of honor, next to one

---

[2] HEBREW: to cause astonishment and awe (Brown-Driver-Briggs' Hebrew definitions).

of the greatest scientists of all time, Isaac Newton. The sum of his life's work was to demonstrate that man came into existence by a fluke of nature and that, at death, a man passed into nothingness. Consequently, the idea that there was a purpose to life was rendered meaningless.

As mentioned, the Darwinian theory of evolution operates by chance, is random, unguided, blind, and purposeless. The creation of life or the intelligent design of life, on the other hand, is deliberately designed, planned, engineered, revealed, and purposeful. Evolution solves problems strictly through a trial and error process. Intelligent design solves problems through a deliberate design process. Present-day engineers also solve problems through a design process accompanied by research and development. It is interesting to note that evolutionists describe "biology as the study of complicated [living] things that give the ***appearance of having been designed for a purpose***."[3] [emphasis added]

Most books about evolution are written by biologists, biochemists, geneticists, medical doctors, or science philosophers. Occasionally, articles appear by mathematicians and even lawyers. However, insights provided by engineers are rare. This is regrettable. I am a pilot and an engineer, having spent fourteen years in the US Air Force, thirty-two years with Hewlett-Packard, Co., and two years with Intel-Micron Flash. Engineers, and pilots, understand the problems associated with designing complicated systems that must work in their targeted environment. The hope is

---

[3] Dawkins, The Blind Watchmaker, 1996, p. 1; Richard Dawkins is an English evolutionary biologist and prolific author. He taught at Oxford. Dawkins is "known as an outspoken atheist. He is well known for his criticism of creationism and intelligent design." (https://en.wikipedia.org/wiki/Richard_Dawkins) His books The Blind Watchmaker and The God Delusion are two of his best-known books. Dawkins will be quoted here extensively.

that this book may provide some valuable insights into what can and cannot be accomplished by a blind, trial and error, pseudo-design process.

This book will distinguish between what has been referred to as **microevolution** and **macroevolution**. Microevolution serves to adapt organisms to their environment and may produce new species, maybe even a new genus. Conversely, macroevolution must create novel new life-forms with new body plans. Evolutionists presume that if microevolution works, then macroevolution must likewise work. For example, "the fact remains that if you accept microevolution, you get macroevolution for free." [4] This idea is wrong. **Microevolution works predominantly through the loss of genetic information or devolution. [5] Macroevolution, on the other hand, requires**

---

[4] Zimmer, Evolution, p. 325; Carl Zimmer is an author and former senior editor for the Discover magazine. He writes a regular column in the magazine, Natural History, concerning evolution. He wrote the book, Evolution: The Triumph of an Idea – a companion book to the seven-part PBS television series, Evolution.

[5] Behe, Darwin Devolves, 2017; Michael J. Behe is a professor of biochemistry at Lehigh University and is a Senior Fellow at the Discovery Institute's Center for Science and Culture. He is best known as an advocate of intelligent design and as the creator of the idea of irreducible complexity which we will address later. His three primary books are must reads regarding arguments against evolution. The first book, Darwin's Black Box: The Biochemical Challenge to Evolution (1996), introduces the idea of irreducible complexity. The second book, The Edge of Evolution: The Search for the Limits of Darwinism (2007), examines what is known about random mutations and their effects on organisms. His latest book, Darwin Devolves: The New Science About DNA That Challenges Evolution (2019), shows that the vast majority of incidents of adaptation are due to mutations of an organism's existing genome that cause the permanent loss of information, not the increase of information as evolutionists would have us think.

complicated and precisely designed new genetic information. Darwinian evolution cannot perform this feat. "There are **no examples of mutations that add any information to the genome**."[6] [emphasis in the original]

The following text will discuss the general topics of molecular biology, how evolution is supposed to work, how evolution works in practice, Intelligent Design (ID), special issues posing significant roadblocks to the theory of evolution, the origin of life problem, the Creation, the Fall, Christ's Atonement, Anti-Christs, and thoughts on the pre-Garden creation. Specific topics include DNA, DNA replication and editing, proteins, protein folding, mutations, and examples of how evolution works. Examples include polar bears, Darwin's finches, drosophila (a fruit fly), E. coli (a bacterium), and malaria (a parasite). Special issues include the Cambrian Explosion, the results of a recent extensive genetic study, the concept of irreducible complexity (using the human eye), and the origin of life problem.

As you study each of these ideas, ask yourself, 'could a random, unguided, blind, and purposeless trial and error process produce this?' or 'does the existence of an Intelligent Designer make more sense?'

---

[6] Spetner, Not By Chance!, p. 173; Lee M. Spetner received his Ph.D. in physics from MIT in 1950 and worked at the Applied Physics Laboratory of Johns Hopkins University for nearly 20 years. Since 1964, **his hobby has been studying organic evolution. For about 10 years, Spetner taught information and communications theory** at Johns Hopkins University. His book Not By Chance! – Shattering the Modern Theory of Evolution (1998) is a must-read regarding evolution and mathematics.

NOTES:
The ~ symbol means 'approximately.'
K = thousand (1,000).
M = million (1,000,000).
B = billion (1,000,000,000).
T = trillion (1,000,000,000,000).

Examples:

~1T means approximately 1 trillion.

~4.6 B means approximately 4.6 billion.

(There is a space added before a B since it can look like an 8.)

Emphasized text looks like this (bold).
Especially emphasized text looks like this (bold, italics).

Sometimes 'earth' is capitalized as 'Earth' when Earth is used as a proper noun.

The first time a reference is given, the full reference is used (author's full name, full title, publisher, year, page). Subsequent references to that reference will include only the author's last name, the main title, and the page.

Thesis Statement

Nearly 200 years after Joseph Smith and Charles Darwin formed their belief systems, there is still a dichotomy between Joseph Smith's God and Charles Darwin's god. The "real conflict ... is not science vs. religion. It is theism versus atheism, ..."[7] The dividing line is which ***worldview*** one adheres to. "The Darwinian worldview [is] that slow, gradual cumulative selection is the

[7] Lennox, God and Stephen Hawking, 2011, p. 12; John C. Lennox wrote a reply to late Stephen Hawking's last book, The Grand Design (2012). His reply is entitled God and Stephen Hawking: Whose Design is it Anyway? (2011). Lennox is an English mathematician, science philosopher and Christian apologist. He is presently Emeritus Professor of Mathematics at the University of Oxford.

ultimate explanation of our existence."[8] [emphasis added] The Lord's ultimate explanation of our existence is found in Moses 1:39, ***"For behold, this is my work and my glory—to bring to pass the immortality and eternal life of man."*** As an active member of the Church of Jesus Christ of Latter-day Saints, I believe in the God of Joseph Smith and that God-the-Father and His son, Jesus Christ, are the creators and intelligent designers of both life and the universe which we inhabit. My experiences as a pilot, physicist, and engineer reinforce that belief. The purpose of this book is two-fold. First, to demonstrate that Darwinian evolution can do no more than make small adaptations to pre-existing organisms (microevolution) usually through devolution (losing genetic information). The ability of life-forms to adapt was designed by God in the beginning. Evolution utterly fails in its ability to explain the creation of new organisms (macroevolution). Second, while leading evolutionists rarely admit their philosophical biases, this book will document that many are, in fact, adamantly anti-Christian and atheistic, meeting all the criteria of modern Anti-Christs. "The very persons who insist upon keeping religion and science separate are eager to use their science as a basis for pronouncements about religion."[9] Two quotes by Richard Dawkins serve as first examples of one preeminent evolutionist's hostility towards religion. Dawkins said, "Darwin made it possible to be an intellectually fulfilled atheist"[10] and "**it is absolutely safe to say that, if you meet somebody who**

---

[8] Dawkins, The Blind Watchmaker, p. 318

[9] Johnson, Darwin On Trial, 1993, p. 8; Phillip E. Johnson recently passed away. He was a Professor of Law at the University of California, Berkeley. He was a critic of evolution and is considered to be the father of the intelligent design initiative. His book, Darwin on Trial, 2nd Edition (1993) analyzed the logic of evolutionist's arguments and their underlying assumptions.

[10] Dawkins, The Blind Watchmaker, p. 6

claims not to believe in evolution, that person is ignorant, stupid, or insane (or wicked)."[11]

Richard Dawkins is also the person who said that evolution is ***"able to derive anything from anything else***. ... ***if we have sufficient time*** ... and ***if there is a mechanism for guiding each step in some particular direction."*** [12] This book is designed to demonstrate that evolutionists who believe in evolution's ability to create 'anything from anything else' are the ones who are blind and irrational because of their absolute belief in Darwinism.

Outline

In Darwin's day, scientists had no concept of the molecular basis of life and its variation. Shortly after Darwin published On the Origin of the Species, Mendel discovered the first principles of genetics in the 1860s. Unfortunately, Mendel's work was largely unknown for almost a century. In 1942, Mendel's theory of genetics was molded into Darwin's theory of biological evolution. The merged theory is referred to as the modern synthesis or Neo-Darwinism. To understand the arguments for or against organic evolution, one must understand the fundamentals of genetics. To understand the fundamentals of genetics, one must first understand the basics of molecular biology.

This book can be divided into three parts. Chapter 2 makes up the first part. This is where you will learn the essentials of molecular biology and genetics. It serves as the basis for the second part of the book. While some concepts can be difficult, every effort has been made to make those concepts understandable. A Glossary of terms is included at the back of the book. The Glossary is structured, i.e., it is not alphabetic. Instead, it is structured as an indented list aka a mind map. Similar ideas are grouped together.

---

[11] Dawkins, *Put Your Money on Evolution*. The New York Times Review of Books: 1989-04-09, p. 35

[12] Dawkins, The Blind Watchmaker, pp. 317-318

This structure should greatly aid understanding. It may be a good idea to read through the Glossary BEFORE beginning Chapter 2.

The second part discusses the details of evolution. Chapter 3 discusses the theory of evolution, how it is purported to work. While the idea of genetic mutation was introduced in Chapter 2, Chapter 3 demonstrates that mutations are the basis for evolutionary change. Detailed examples of mutation and their probability are given in Chapter 4. Chapter 5 discusses how engineering works in contrast to evolution. Engineering is also presented as a rough substitute for Intelligent Design. Unfortunately for evolutionists, evolution does not work as theorized. It works predominantly through devolution. Devolution means the loss of genetic information versus evolutionary theory's presumed accumulation of new, useful information. Detailed examples of devolution are given in Chapter 6. This is followed by several key topics are in Chapter 7.

The examples of devolution in Chapter 6 include polar bears, Darwin's finches, Drosophila (fruit fly), E. Coli (bacterium), and Malaria (parasite). The polar bear example is relatively simple and sets the tone for the chapter. Darwin's finches were supposed to be a wonderful example of evolution shortly after Darwin published his theory. It turns out that they were, instead, an extremely poor example. Drosophila is important since it was the first organism whose entire genome was examined. The research goal was to discover how macroevolution occurred. The results were disappointing. The experiment involving the E. Coli bacterium is the longest-running evolutionary experiment in history. While the bacterium changed, it was through devolution. Despite tens of thousands of generations and billions of bacteria, E. Coli is still E. Coli. The discussion about malaria paints a picture of what the limits of evolution are. Part 2 addresses Thesis 1 which states that evolution is incapable of macroevolution.

The topics in Chapter 7 were selected carefully. They include the Cambrian Explosion, a recent report on a genetic study of millions of organisms, the concept of irreducible complexity using the human eye as an example, and the origin of life. The Cambrian Explosion is the preeminent example of major macroevolutionary steps in too short a time to be explained by Neo-Darwinian evolution. The genetic study is expansive and its findings contrary to evolutionary tenets. The evolution of or the design of the mammalian eye has been a point of contention for two hundred years. Evolutionists use presumed defects in the structure of the eye bolstering their idea that, while evolution can 'design' some amazing structures, it is not perfect and makes mistakes. I will show that every presumed design flaw is instead a brilliant design feature – the design IS perfect for the task! Evolution cannot explain the macroevolution of living organisms, let alone the origin of life itself. Evolutionary ideas are used to propose scenarios for how life originated. These scenarios will be shown to be more fantasy than fact. A short section discussing human evolution and mankind's presumed tree of life is included.

Part three departs from the details of evolution. Instead, Chapter 8 focuses on an alternative to naturalistic evolution – Intelligent Design. Nature does not just appear to be designed; it is designed! Chapter 9 addresses Thesis #2, modern-day Anti-Christs. Many of the most outspoken evangelists for evolution meet all the requirements of being Anti-Christs. Numerous quotes are provided to establish the case that many are, indeed, Anti-Christs. Chapter 10 discusses the Lord's doctrines of the Plan of Salvation, including the Creation, the Fall, and Christ's Atonement. The Plan of Salvation was purposeful; every contingency was planned for. The Creation, the Fall, and the Atonement of Jesus Christ are intimately intertwined. These three pivotal events cannot stand alone. For me, I cannot phantom how evolution could be part of the Creation. This chapter also discusses topics regarding the Earth's creation, such as being created near Kolob in a terrestrial state. According to early leaders of the church who were personal acquaintances with the Prophet Joseph Smith, it was not until after Adam's and Eve's

Fall that the Earth fell through space and ended up in our current telestial solar system.

There are two appendices. The Church of Jesus Christ of Latter-day Saints has issued two Doctrinal Expositions by the First Presidency of the Church regarding the *Origin of Man*. The first in 1909 carries the signature of President Joseph F. Smith. The second in 1925, carries the signature of President Heber J. Grant and is a word-for-word extract from the 1909 Exposition. The complete 1909 and 1925 Doctrinal Expositions are given in Appendix 1. Appendix 2 is the author's critique of a display on evolution at the Monte L. Bean Museum at Brigham Young University.

When you have finished this book, you should have a sound understanding of how evolutionary theory is supposed to work and how it actually works, with examples. You should be familiar with the immense improbabilities involved in the chance-based creation of living organisms. You should be in awe of molecular machines such as DNA, replisomes, spliceosomes, chaperonins, proteins, and their intricate shapes which make them functional and versatile. You will understand the need for an Intelligent Designer. You will also be aware of the atheistic worldview of many evolutionary evangelists and understand that their arguments regarding evolution are not intended to further the eternal progress of mankind. Finally, you will know Church doctrine regarding the Creation, the Fall, and Christ's Atonement and that these three critical events rise and fall together. The Plan of Salvation puts limits on the learnings of science and scientists' understanding of the Divine.

2. INTRODUCTION TO MOLECULAR BIOLOGY, THE KEY TO CHALLENGING EVOLUTION

Darwin developed his theory of evolution based on his observations of life using the naked eye. Variations between individual life-forms were essential to his theory. He had no idea what the source of these variations was. It was nearly 120 years later, with the discovery of DNA, that evolutionists could begin to understand the source of variation – variation that occurred at the MOLECULAR level! The principles of evolutionary theory are deceptively simple at the naked eye level. Unfortunately, at the molecular level, understanding the mechanics of variation and its effects on an organism gets much more difficult. It is necessary to discuss the fundamentals of molecular biology preparatory to understanding the detailed mechanisms of evolution. For a more thorough but still simplified overview of molecular biology, please see Clark and Russell's Molecular Biology – Made Simple and Fun (4th Edition, Cache River Press, 2010) and Meyer's Signature in the Cell: DNA and the Evidence for Intelligence Design (Harper One, 2010). This discussion will involve a lot of new terms and names of molecules and biological structures. Try to focus on the fundamentals while marveling at how awe-inspiring everything in our cells works. The cell and all its contents are indeed 'wise and wonderful.'

The following sections will discuss the fundamentals of DNA and biological information storage, how the information is encoded in codons, how DNA is replicated inside the cell, how DNA is translated into mRNA, how proteins are formed, and how proteins fold up into elaborate 3-D structures. DNA and proteins are essential to life. To put things in perspective, the number of DNA molecules, genes, and proteins that are found in every individual cell will be discussed. That discussion will also include the concept of a minimal cell – what is the minimum number of genes that are required for the smallest cell to exist? It will become obvious that a minimum cell is not the same as a trivial cell. Even the simplest cell is extraordinarily complex. Ask yourself throughout this discussion if a blind, random, naturalistic, purposeless, trial and error process could have generated such precise, watch-like biological machines?

There are many new terms and some concepts may seem difficult at first. Have faith – you can do it! I have made every effort to make the concepts understandable to the layperson. A Glossary is provided at the back of the text to help with the vocabulary and the basic ideas. **Remember, understanding molecular biology is the key to understanding how evolution is supposed to work and why it cannot work that way**.

DNA

All life-forms require four macromolecules (large molecules) to function. These macromolecules are nucleic acids (DNA, RNA), proteins, lipids (fats), and carbohydrates (sugars). Only the first two are pertinent for our present purposes. First, we will discuss DNA. This will be followed by a discussion of how DNA encodes proteins. Finally, we will talk about proteins, how they fold into complicated three-dimensional (3D) shapes, and the purposes for their shapes. Note that DNA is an abbreviation for **d**eoxyribo**n**ucleic **a**cid; RNA is short for **r**ibo**n**ucleic **a**cid. The ribose sugar molecule forms the backbone of the DNA and RNA molecules. The ribose molecule in DNA ($C_5H_{10}O_4$) has one less oxygen molecule than the sugar molecule in the RNA molecule ($C_5H_{10}O_5$), hence the term 'deoxy' in the name for DNA.

In the early 1950s, the details of cell structure, DNA, and proteins started to be discovered. DNA is the well-known double-helix molecule that contains the genetic information found in all living organisms. See Figure 1. DNA has two helical backbones connected by pairs of 4 chemical bases. The bases are referred to as G, C, A, and T (Guanine, Cytosine, Adenine, and Thymine respectively). G always pairs with C and A always pairs with T. The sequence of base pairs forms an extremely long sequence of genetic data. Each side of the double helix encodes the same information, only in the reverse order. Operationally, these base sequences are organized into groups of 3 sequential bases, called codons. Two examples are GAC and TGG. Since there are 4 bases taken 3 at a time, the number of permutations[13] that can be coded by a codon equals 4^3 or 64. The mathematical expression 4^3 means 4 taken to the power of or exponent of 3, i.e., $4 \times 4 \times 4 = 4^3 = 64$. This exponential notation will be used throughout the book.

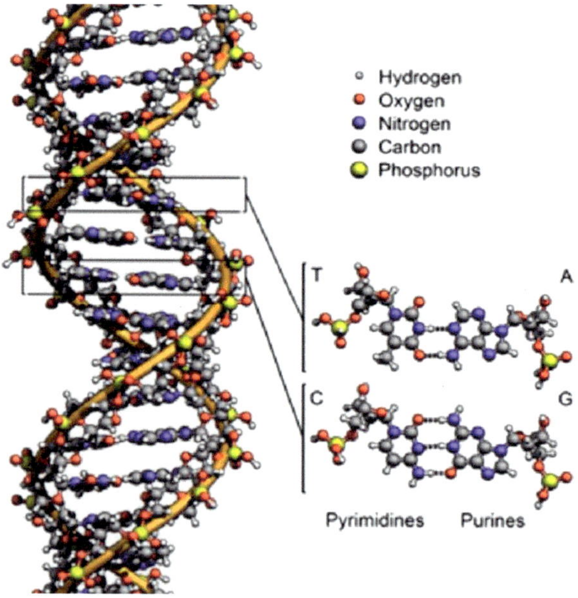

FIGURE 1. THE STRUCTURE OF DNA.[14]

---

[13] Permutations are arrangement of things in which order matters. For combinations, order does not matter. For example, the combinations of CAT, ATC, TCA, etc. are considered the same while GAT would be different. The permutations CAT, ATC, TCA, etc. are considered different. Arranged as a codon, the base strings CAT, ATC, TCA, etc. provide a unique code which selects a specific amino acid.

[14] By Zephyris - Own work, CC BY-SA 3.0, https://commons.wikimedia.org/w/index.php?curid=14709798

Why is a codon composed of 3 bases? We will see shortly that one of the things that DNA must do is code for proteins. Proteins are required for life. Twenty standard amino acids are used to build proteins. Amino acids are small organic compounds with two standardized ends (an amine group and a carboxyl group) and a side chain. See Figure 22. Each of the 20 amino acids has a unique side chain. If a codon was composed of only 1 base, it could only code for $4^1 = 4$ different amino acids. If a codon had 2 bases, it could code from $4^2 = 16$ amino acids. Consequently, 3 bases are required to select from 20 standard amino acids ($4^3 = 64 \geq 20$).

In cells without a nucleus (prokaryotes), DNA is found floating in the cytoplasm (the fluid inside the cell's membrane). In cells with a nucleus (eukaryotes), DNA is found primarily in the nucleus, though some forms of DNA are found in another cell organelle called the mitochondria. Most living organisms are prokaryotes. All prokaryote life-forms are unicellular (single-celled). While many eukaryote life-forms are also unicellular, eukaryote cells are the basis for all more complicated life-forms, including humans.

The definition of genes, chromosomes, and genomes will prove valuable as we delve deeper into cell chemistry and function. A region of DNA that performs a single function is called a gene. A long strand of DNA containing many genes is called a chromosome. The total genetic information stored in an organism's genes, which may include multiple chromosomes, is called the organism's genome. In all eukaryote cells, multiple genes are assembled into chromosomes. Not all genes do the same thing.

Genes perform three different but hierarchical functions: structural, operational, and regulatory functions. Regulator genes turn operator genes on or off. Operator genes then govern the transcription of structural genes. Finally, structural genes are used to build proteins. Variations or mutations in each type of gene will have different effects on an organism. Variations in structural genes are the easiest to understand and will be our primary focus. Unfortunately, the functions of regulatory and operational genes are extremely complicated and most are not understood. Mutations in these two types of genes can cause huge changes and are usually devastating to the organism. This will become evident when we discuss mutations in drosophila later.

Here are some interesting facts about human DNA. The human genome comprises ~3 B base pairs. A base pair is either T-A or C-G. Each human's DNA is ~99.9% identical to every other human's DNA. Consequently, ~0.1% of our genome accounts for all human differences. Using only the four letters representing the nucleotide bases, T, A, C, and G, one could type out the human's genome. Typing at a typical 40 words per minute, each word averaging 5 characters, 8 hours per day, it would take ~90 years to complete. Since every cell contains DNA, and because there are ~30T cells in our bodies, our bodies have an enormous amount of DNA. "If you put all the DNA molecules in your body end to end, the DNA would reach from the Earth to the Sun ... over [300] times." [15] For reference, the distance from the Earth to the Sun is ~93M miles (1.5 x 10^{11} meters) while each base pair is about 0.5 nm (5 x 10^{-10} meters) long. That is an enormous number of base pairs!

DNA Replication (Transcription)

The purpose of DNA is to store a cell's genetic information and to produce proteins that are the workhorses of cell life. Every time a cell divides, it is necessary to replicate the original genome – its chromosomes and genes. All living organisms have multiprotein machinery that replicates DNA's information. The components of the multiprotein machines self-assemble on the DNA molecule. Self-assembly means that the necessary molecules (proteins and specialized RNA) are floating in the cell's cytoplasm. Since these molecules are floating and moving rapidly in the small confines of

---

[15] Anne Marie Helmenstine, *10 Interesting DNA Facts*, ThoughtCo., https://www.thoughtco.com/interesting-dna-facts-608188

the cell, they quickly find their ways to appropriate attachment points on the DNA. When all the necessary parts are assembled, the molecular machine can begin to operate. Basically, DNA needs to be unwound, copied, and then rewound. One of the primary copying machines is called a replisome.

The first step in DNA replication (copying) is to unwind the DNA into two single strands. Each strand is then copied exactly after which the newly replicated strands combine to form a new DNA molecule identical to the original DNA. The purpose of the replisome is to precisely copy the original DNA. Replisomes are extremely complicated assemblies, composed of two molecular complexes.[16] One replisome complex synthesizes the leading or coding DNA strand in primary order, slightly ahead of the lagging strand. The other complex synthesizes the lagging or template strand (lagging because it is synthesized slightly behind the leading strand). The lagging strand is the complement or reverse of the leading strand. Since DNA can only be assembled in one direction, the primary direction, the lagging strand presents a unique problem – it is backward. To solve this problem, the replisome first passes a section of the lagging strand through, snips the strand section off from the full lagging strand, and then stops the lagging strand. The strand section is then replicated in reverse order (the primary order). The replicated section is then turned around and attached to the previously copied lagging strand section. Consequently, the lagging strand is built section by section in reverse order. Meanwhile, the leading strand waits for each lagging strand section to be replicated before it continues being replicated.

It is one thing for a biological system to 'figure out' how to copy the leading or coding strand in order as it comes. It is quite another to 'know' that the lagging strand is backward, that it needs to be copied backward, that it needs to been broken into sequential

[16] https://en.wikipedia.org/wiki/Replisome

snippets for ease of handling, and that **it then needs to turn the snippets around** to be attached to the growing copy of the lagging strand. Think about how this could have happened by chance!

The replisome is composed of the following protein subcomponents: helicase, RFC, PCNA, gyrase/topoisomerase, SSB/RPA, primase, ligase, DNA polymerase III, and RNAse. Do not worry about the names of the individual components, rather pay attention to the facts that there are many of them, that they all must be present, and that they must all work precisely together to replicate DNA, i.e., it is irreducibly complex.

The replisome uses several mechanisms to ensure DNA replication accuracy. The first mechanism is the ability to accurately copy long stretches of DNA. The second is to catch its own mistakes as they happen and repair them immediately – a type of proofreading. A third is to constantly monitor the DNA for damage – certain proteins detect the damage and tell the cell to stop replication until it can be repaired.[17] The result is that DNA replication is extraordinarily accurate. "In the initial pairing of bases with the template DNA strand, there is around one error for every 100,000 nucleotides paired. Polymerase enzymes (enzymes that bring about the formation of particular polymers, especially DNA or RNA) proofread the new strand of DNA against the template strand and fix errors. This fixing reduces errors to around one error for every 10 billion nucleotides, an extremely accurate process. The one in 10 billion errors occur when a polymerase incorrectly replaces the original nucleotide with another incorrect nucleotide. These rare errors are the **cause of genetic mutations** and cancer."[18] [emphasis added] It appears that these three processes were

---

[17] https://education.seattlepi.com/mechanism-ensures-accuracy-dna-replication-4818.html

[18] https://basicbiology.net/micro/genetics/dna-replication

designed to ensure the integrity of an organism's genome, i.e., to prevent copying errors or mutations.

Using our typing analogy, our hypothetical typist types about 40 words per minute or about 200 characters per minute and types continuously for 8 hours per day without weekends or holidays. We will let one typing character represent one nucleotide base. The average typist has an error rate of about 2%, or 2 out of a 100. Therefore, in one minute, an average typist makes 4 errors. That means that our average typist makes an error every 15 seconds. **To match the 1 in 10 B error rate of a replisome, the average typist would have to type for almost 300 years before making their first mistake!** Consequently, evolution has a scarcity of mutations to bring about evolutionary change. Try to imagine how replisomes or polymerases could have arisen by chance.

Evolution demands genetic variation or mutations to work. Healthy biological diversity requires some small level of genetic variation. However, living organisms necessarily must minimize genetic errors to ensure long-term survival. There are numerous types of copying errors. Simple types of DNA replication errors include base substitutions, insertions, and deletions. Errors in the genetic code can have three effects. They may not affect the organism at all, i.e., they are neutral mutations, they can be harmful, or they can be beneficial. **Most errors are neutral, many are harmful, and very few are beneficial.** This is because life's molecular machines are already finely tuned for nearly perfect operation, like the replisome.

The replisome is an example of an extremely complicated biomolecular system or biomachine. Humans cannot see the replisome in action; molecules are smaller than the wavelength of light and, thus, are not visible to the human eye even with the most powerful optical microscopes. Fortunately, using numerous modern imaging techniques, the details of how the replisome functions have been discovered. Dr. Drew Berry, one of the world's foremost animators of biological molecular machines, creates amazing animations that "reveal the microscopic worlds inside our bodies." He has created an animation showing how the replisome works.

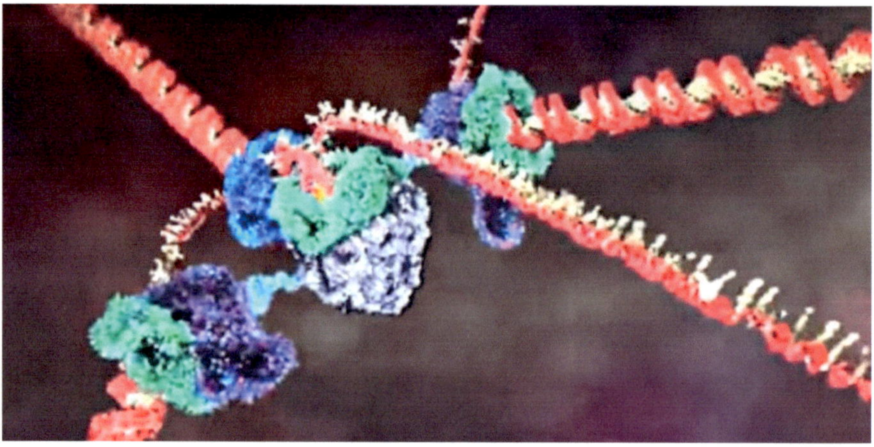

FIGURE 2. A FRAME-GRAB OF A DREW BERRY ANIMATION DEPICTING THE REPLISOME IN ACTION. [19]

Replisome Animation

Dr. Drew Berry is the Biomedical Animations Manager at the Walter+Eliza Hall Institute of Medical Research.[20] On the WEHI webpage is an animation entitled *Animations of Unseeable Biology* See Figure 2. The first part of the animation (2:44-4:25) shows some components of the replisome in action. The entire animation inspires wonder and awe. Berry calls these biological systems "astonishing molecular machines." Later in the animation, Berry refers to the cell's systems as "molecular clockwork." It should be noted, that to make the basic processes more visible, he left out "error correction and a bunch of other things." Consequently, while

---

[19] TEDxSydney, https://www.ted.com/talks/drew_berry_animations_of_unseeable_biology; DNA Replication by Drew Berry, The Walter and Eliza Hall Institute (by permission)

[20] https://www.wehi.edu.au/people/drew-berry

the animation is amazingly complicated, it is much simpler than the complete process.

The pieces of the replisome (most of them proteins) are in constant motion. When proteins are formed, the various amino acid sequences fold into exquisite 3-dimensional (3D) shapes. The shapes are like tinker toys with specialized lock and key areas on their surfaces. These lock and key areas are exactly shaped to accept the key and lock surfaces of other proteins; these combined proteins are designed to perform specific functions. Often, when one protein attaches to another protein, it changes the original protein's shape into a secondary functional shape. Each movement of the replisome is caused by an attached protein. When the attached protein subsequently detaches, the replisome protein returns to its original shape, and the sequence repeats. This molecular clockwork truly is an "astonishing molecular machine." Remember that the final replication error rate is ~1 error for every 10 B base pair replications.

DNA Translation (Forming Proteins)

One of the most important tasks for DNA is to create proteins. Life would not be possible without proteins. Proteins are used everywhere to build, operate, and maintain living cells. There are, in fact, eight types of proteins: structural, enzymatic, hormonal, defensive, storage, transport, receptor, and contractile. In general, one gene (a single functional sequence of DNA) codes one protein. To create or synthesize a protein, a gene must first be translated from DNA to mRNA (messenger RNA) inside a cell's nucleus. The mRNA then migrates out of the nucleus into a ribosome (a cellular organelle) where the creation of the final functional protein is completed.

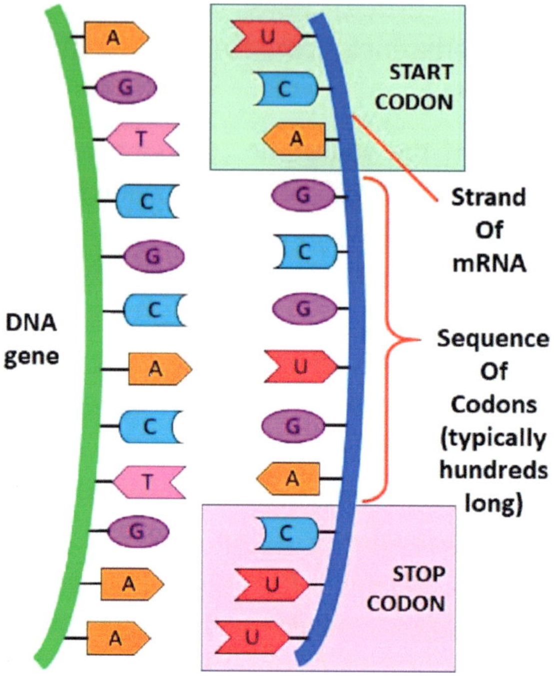

FIGURE 3. A DIAGRAM OF MRNA BEING CREATED FROM DNA. [21]

David Goodsell has stated, "the key molecular process that makes modern life possible is protein synthesis since proteins are used in nearly every aspect of living. **The synthesis of proteins requires a tightly integrated sequence of reactions, most of which are themselves performed by proteins.**"[22] **Code translation from DNA to a completed protein requires hundreds of proteins.**[23] This situation presents an interesting paradox, where did the proteins come from that are necessary to make proteins? David Goodsell has conjectured, this paradox "is one of the unanswered riddles of

---

[21] D. John Redman

[22] Goodsell, The Machinery of Life, p. 41; Meyer, Signature in the Cell, p. 133; David Goodsell is a structural biologist who works at the Scripps Research Institute. He is best known for his amazing watercolor paintings of cellular components. You can see his work in this book in Figures 5, 6, and 8.

[23] Meyer, Signature in the Cell, p. 134

biochemistry: which came first, proteins or protein synthesis?"[24] Of course, they might have been intelligently created simultaneously for that very purpose. This has important implications regarding the origin of life. For now, note again how complicated and amazing the translation process is that converts DNA information into proteins.

During transcription or replication of DNA, a second copy of two-stranded DNA is created. During translation (the translation of DNA language into protein language) the DNA code is copied to a single strand of RNA and edited preparatory to producing a protein. Figure 3 depicts one-half of the DNA on the left-hand side. On the right-hand side is a strand of mRNA which has the correctly paired base molecules. Cell machinery reads the RNA code in groups of three bases called codons. Up to 98% of human DNA is non-coding. This does not mean that 98% of human DNA is unused, it is just not used for coding proteins. These non-coding sections of DNA are called introns. The coding segments are called exons. It is necessary to remove the introns from the RNA before the resulting mRNA (messenger RNA) leaves the nucleus on its way to a ribosome where proteins are made. See Figure 4.

The top row (a) of Figure 4 depicts the original DNA with exons and introns. The leading end of exons (on the right of the exons in the diagram) are annotated with a 5′ while the trailing end of exons (on the left end in the diagram) are annotated with a 3′. There is a string of five carbon atoms in the deoxyribose sugar backbone of the DNA (deoxyribose, $C_5H_{10}O_4$). If one circumnavigates the carbon string in a clockwise manner, 3′ refers to the third carbon atom and 5′ refers to the fifth carbon atom in the string. The DNA molecule always grows from the 5′ end.

---

[24] Goodsell, The Machinery of Life, p. 45; Meyer, Signature in the Cell, p. 134

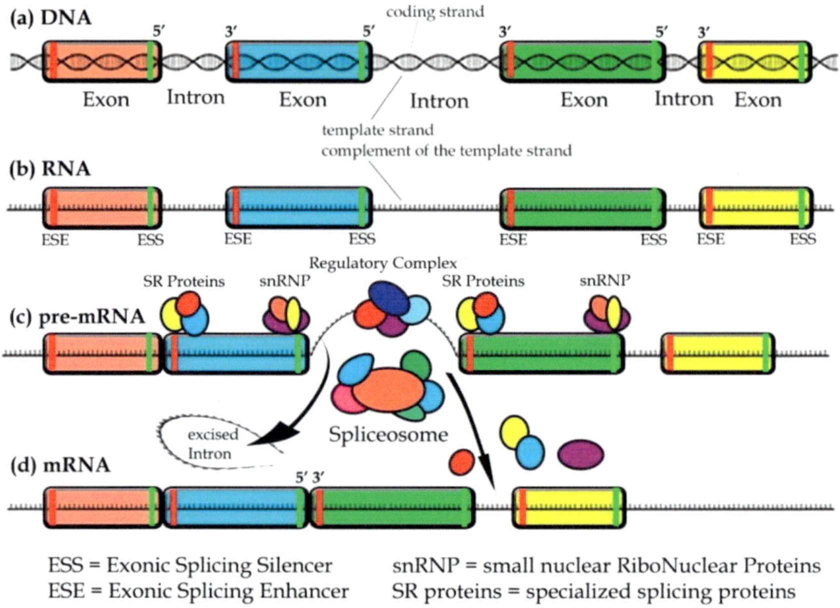

FIGURE 4. AN OVERVIEW OF HOW DNA IS TRANSLATED INTO MRNA.[25]

The second row (b) depicts the single-strand RNA string copied from the DNA. Notice the start/stop identifiers have been copied from the DNA. Exons have a start code at their beginning referred to as an ESS (Exonic Splicing Enhancer) while there is a similar stop code at the ending called an ESE (Exonic Splicing Silencer). The ESS and ESE indicators tell the cutting and splicing machinery where to cut and splice.

The third row (c) shows the self-assembled start/stop identifiers (snRNP and SR proteins), the regulatory complex (the cutter), and the spliceosome (splicer) on the RNA string. The molecular components required to perform cutting and splicing are present inside the nucleus, floating around in the cytoplasm. Self-assembly means that the necessary proteins and other molecules quickly bump into their respective destinations and attach to them.

---

[25] D. John Redman

They attach to the RNA in highly specific ways forming incredibly tiny biological machines of various sizes, shapes, and complexities.

The fourth row (d) shows the newly spliced RNA without any introns after the identifiers. The regulatory complex and the spliceosome have self-disassembled. After splicing is completed, the various assemblies disassemble and begin to float around quickly attaching to the next exon. Like the replisome, the spliceosome is an extremely complicated micromachine designed for a purpose.

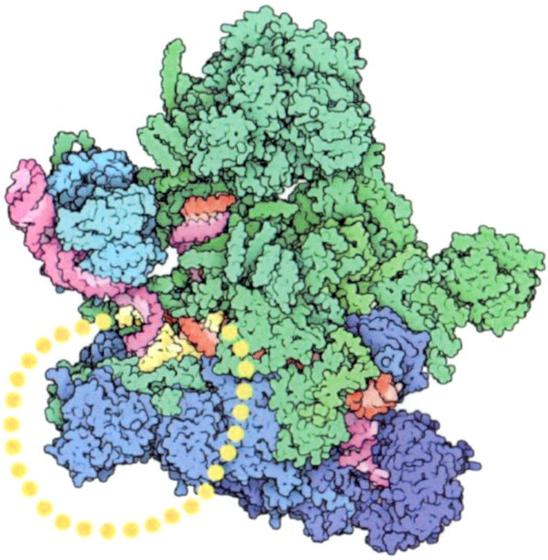

FIGURE 5. A MODEL OF A YEAST SPLICEOSOME. [26]

The spliceosome is a large and intricate molecular machine found in the nucleus of eukaryotes. The yeast spliceosome sequence and its 3D structure have previously been discovered and made public. A model of a yeast spliceosome is shown in Figure 5. The yeast spliceosome (Spp42) has 10,574 amino acids from 37

---

[26] Protein model by David Goodsell, Protein of the month, http://pdb101.rcsb.org/motm/245

proteins, 3 snRNA molecules, some other molecules, and a combined molecular mass of ~1.3MDa.[27]

Often, biochemists want to know how many atoms make up a molecule. Since there are over 100 types of atoms in the Periodic Table, they developed an average mass unit called a Dalton (Da). By definition, 1 Da is equal to 1/12 of the mass of a carbon-12 atom. The Da unit is named after John Dalton who proposed the atomic theory of chemistry around 1808.[28]

In mid-2017, the first atomic structure of a human spliceosome was announced. "The final atomic model of the human [spliceosome] contains 14,496 [amino acids] from 46 proteins and 380 RNA nucleotides, which amount to approximately 1.7 mega-Daltons."[29] For comparison, the average human protein is ~40KDa. At ~1.3MDa and ~1.7MDa, the yeast and human spliceosomes are monster molecules.

Every single human cell contains ~100K spliceosomes which are responsible for removing more than 200K different intron sequences and splicing the remaining genetic code back together before being translated into proteins. To view a 3D animation of RNA Splicing (1:37), go to the DNA Learning Center website sponsored by the Cold Spring Harbor Laboratory.[30]

It can readily be observed that the spliceosome, like the replisome, is an irreducibly complex system. An irreducibly complex system is one that cannot work unless every individual

---

[27] Yan, Hang, Wan, Huang, Wong, Shi, *Structure of a yeast spliceosome at 3.6-angstrom resolution*, Science, 11 Sep 2015, https://science.sciencemag.org/content/349/6253/1182?ve=

[28] https://en.wikipedia.org/wiki/John_Dalton

[29] Zhang et al, *An Atomic Structure of the Human Spliceosome*, Cell, Vol. 169, Issue 5, 18 May 2017, pp. 918-924

[30] Cold Spring Harbor Laboratory, DNA Learning Center, https://dnalc.cshl.edu/view/16938-3d-animation-of-rna-splicing.html

component is fully functioning and cooperating with the other components to complete a task. Additionally, an irreducibly complex system realistically cannot work as it is being developed, otherwise, all of the other cooperating components would have to change in perfect harmony if even one individual component were to change. As before, ask yourself if such a system could have come about through a time-consuming, error-fraught, blind, trial and error process based upon random mutations? One can hardly think of a single biological system that is not irreducibly complex. Such systems require an intelligent designer.

Protein Folding

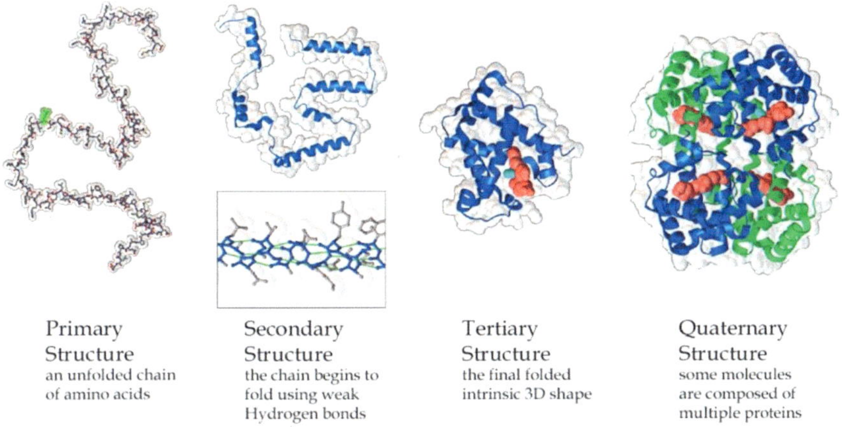

FIGURE 6. THE SEQUENCE OF FOLDING A PROTEIN. [31]

It was mentioned previously that functional proteins must fold into a predetermined shape. "Life as we know it would be bust if

---

[31] Modified by D. John Redman. Illustrations by Jessica May & David S. Goodsell, *What is a Protein?* Protein Data Bank; pdb.101.rcsb.org; Animation at http://pdb101.rcsb.org/learn/videos/what-is-a-protein-video (6:57)

proteins didn't fold into functional 3-D structures."[32] Proteins are manufactured using mRNA (Figures 3 and 4) after the mRNA migrates from the nucleus into a ribosome. Ribosomes are complex biochemical organelles inside cells that translate the information in the mRNA's genetic code into a chain of amino acids. In Figure 6, the left-most illustration depicts the first step in protein creation–an unstable chain of amino acids referred to as the Primary structure. To become biologically functional, the chain of amino acids must undergo various contortions (folding) to reach its native 3-D conformation. Native means the intended, final, stable, and functional 3-D protein shape. Protein folding begins almost as soon as the chain emerges from the ribosome. See the second image from the left in Figure 6. This initial folding is a result of hydrogen bonds being formed between parts of the amino acids. Hydrogen bonds are weak, primarily covalent chemical bonds formed between two atoms that share electrons. This preliminarily folded protein is referred to as its Secondary structure. In most cases, this initial folding is not enough to reach the native state. Cells have other biomachines such as chaperones and enzymes that help the chain fold into its final Tertiary shape, the third image from the left in Figure 6.

The inside of cells is primarily water. Water is composed of two hydrogen atoms and one oxygen atom (H_20). Water is a polar molecule. Polar means electrically charged. Because of chemical bonding between the oxygen and hydrogen atoms, the hydrogen parts of the molecule are positively charged, while the oxygen part of the molecule is negatively charged. The inside of cells is, therefore, a polar environment since cells contain water. Some parts of proteins are hydrophobic (water-hating, non-polar) while other parts of proteins are hydrophilic (water-loving, polar). In

---

[32] Everts, Sarah, *Protein Folding: Much More Intricate Than We Thought*; https://cen.acs.org/content/cen/articles/95/i31/Protein-folding-Much-intricate-thought.html

general, proteins fold so their hydrophobic (water-hating) parts are on the inside while the hydrophilic (water-loving) parts are on the outside interacting with the water. The insides of cells are crowded with macromolecules. Because a folded protein is smaller than an unfolded protein, the crush of macromolecules can encourage premature and incorrect folding. Consequently, certain types of proteins need a temporary safe environment to fold, i.e., an environment away from the polar, crowded cytoplasm.

Some chaperone molecules permit certain proteins to move into their inside volumes where they can be temporarily protected. Chaperone molecules with cavities for proteins to hide are called chaperonins. An example of a chaperonin is found in the E. coli bacterium (a prokaryote). The chaperonin is composed of the GroEL and GroES proteins. See Figure 7. GroEL has been referred to as cylindrical megamachine.[33] These are nearly identical to the proteins Hsp60 and Hsp10 which are found in eukaryotes. These chaperonins are found near the exit of the ribosome. GroEL entices partially folded amino acid chains (secondary structures) into its cylindrical center which is lined with hydrophobic amino acids. Once the chain is safely inside the cylinder, GroES caps GroEL's cylinder. At this point, GroEL undergoes a significant conformational (shape) change. The hydrophobic amino acids inside the cylinder retract making the inside of the cylinder hydrophilic like the inside of the cell. Subsequently, the protein chain can fold up in a safe environment, isolated from the cell's chemically jam-packed environment. Once folding is complete, the protein is released back into the cell's interior.[34]

---

[33] Everts, Sarah, *Protein Folding: Much More Intricate Than We Thought*; https://cen.acs.org/content/cen/articles/95/i31/Protein-folding-Much-intricate-thought.html

[34] https://en.wikipedia.org/wiki/GroEL

Figure 7 shows the GroEL and the GroES proteins. On the left is a top view of GroEL. The protein to be folded hides in the center hydrophobic pocket (the reddish center) protected from exterior water. The right image is a side view of GroEL capped by the GroES protein. The conformational change occurs on the inside of the GroEL protein after GroES attaches.

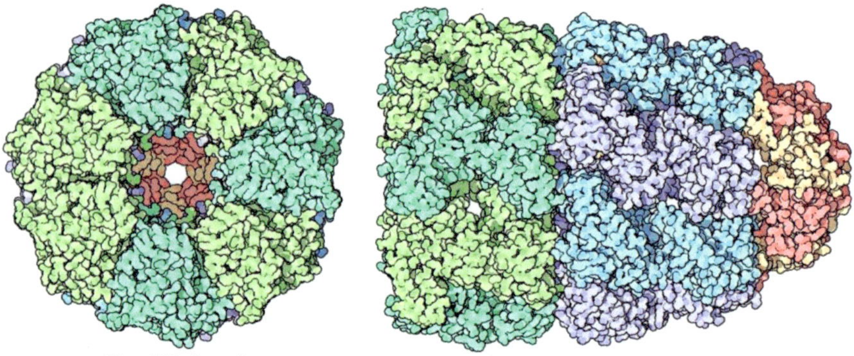

GroEL (top) GroEL (side) with GroES (side)

FIGURE 7. A DEPICTION OF A CHAPERONIN PROTEIN. [35]

Another interesting phenomenon occurs during protein creation and folding. Remember that a codon can code for 64 different amino acids, however, there are only 20 standard amino acids used to create proteins. Consequently, different codons can code for the same amino acid. It has recently been discovered that even though the same amino acid is selected by the different codons, they can have different effects on the creation/folding process besides selecting an amino acid. It turns out that some of the codons cause faster or slower protein creation. Interestingly, slow codons occur between large chains of amino acids that fold largely independently of the subsequent chain. This allows time for the complex folding to occur before moving onto the next chain segment. Not only does this slow down the allowed time for folding

---

[35] PDB-101, Molecule of the Month, *Chaperones*, https://pdb101.rcsb.org/motm/32

to occur, but it also reduces the error rate in chain generation.[36] Amazing!

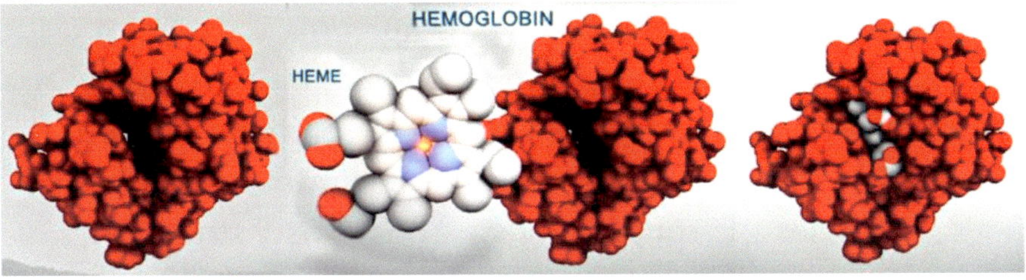

FIGURE 8. THE SHAPE OF HEMOGLOBIN HAS A CAVITY THAT EXACTLY MATCHES THE SHAPE OF HEME. [37]

The final 3D shape of a protein is critical to its function. As we have seen, when other proteins or enzymes attach to the original protein, it may undergo a conformational change. When the proteins or enzymes detach themselves, the protein returns to its original shape. This is what is happening in multiple locations repetitively in the replisome and spliceosome animations. These are dynamic interactions. There are also static interactions. A good example of how 3D shapes statically interact is shown in Figure 8. The left image shows the hemoglobin molecule. The center image shows how the heme molecule fits snugly into the hemoglobin molecule. Heme contains an iron atom at the center which enables it to store oxygen. The right image shows how perfectly the heme molecule fits inside the hemoglobin protein. This fit is extraordinarily precise; could this fit have occurred by chance?

Often, functionally complete proteins will form into Quaternary forms (see Figure 6) for capacity or efficiency reasons.

---

[36] Everts, Sarah, *Protein Folding: Much More Intricate Than We Thought*; https://cen.acs.org/content/cen/articles/95/i31/Protein-folding-Much-intricate-thought.html

[37] Frame grabs from the animation: *What is a Protein?* Protein Data Bank; pdb.101.rcsb.org; Animation at http://pdb101.rcsb.org/learn/videos/what-is-a-protein-video (6:57)

This is true for hemoglobin in which four hemoglobin molecules attach to each other inside the blood cell to carry more oxygen in a smaller space than if the hemoglobin molecules remained separate. While hemoglobin forms a quaternary structure using nearly identical subunits, most quaternary structures are comprised of different subunits. Many proteins require the assembly of subunits before they can begin to work properly.[38] Both replisomes and spliceosomes are quaternary structures.

Examples of Genes and Proteins in Cells

At this point, some examples of how many genes and proteins occur in different kinds of cells should prove interesting. In general, bacteria (prokaryotes) have smaller genes, code fewer types of proteins of smaller average size, and have smaller cell sizes than eukaryote cells. Bacteria have ~3K genes, most of which code proteins. The typical eukaryote cell has ~15K genes while human cells have ~35K-47K genes, depending on how they are counted. The average bacterium has a genome consisting of 1-2M base pairs. Humans have two pairs of 23 chromosomes comprising ~3 B DNA base pairs. The average protein length (number of amino acids) is 267 for bacteria, 361 for eukaryotes, and 375 for humans. Most proteins range in length from 50 to 2000 amino acids. An average bacterial cell contains ~2M proteins while larger human cells contain ~1-3 B proteins of various kinds.[39]

Minimal Cell

Given the above information about living organisms, one might ask, 'What is the minimum number of genes required in a free-living, self-regulating cell?' Scientists led by Craig Venter and

---

[38] Godbey, W. T., *Proteins*, An Introduction to Biotechnology (2014); extract in ScienceDirect, *Protein Quaternary Structure*; https://www.sciencedirect.com/topics/neuroscience/protein-quaternary-structure

[39] https://en.wikipedia.org/wiki/Human_genome

Clyde Hutchinson examined the mycoplasma genitalium bacterium which has the smallest known genome, possessing 517 genes and 581K base pairs. Researchers selectively trimmed the bacterium down to 473 genes. They then proceeded to knock out (remove) each of the 473 genes one-at-a-time. They discovered that 375 genes were required for the modified bacteria to continue to live and reproduce.

Surprisingly, in the world's smallest known bacterium, they could not determine the purpose of 149 of those 375 essential genes (40%) – without their presence, however, the cell could not live. This led the researchers to state, "We've sequenced everything on this planet, and we still don't know 149 genes that are most essential for life!"[40] It is evident that **we have a long ways to go before we know how all genes accomplish their elegant work**. Note that, according to this experiment, **the absolute minimum cell that can live and reproduce requires 375 genes, over 500K base pairs, and has over 1M proteins of various kinds!** This fact presents severe difficulties for evolution and the origin of life.

Pause now and reflect on what we have learned about DNA, information storage, replication, replisomes, spliceosomes, mRNA and protein creation, protein folding, chaperones, multiple codons coding for the same amino acid having different effects, how many genes and proteins are inside cells, and minimal cells. Ask yourself, could a random, unguided, blind, goalless, and purposeless, trial and error development process have produced these extraordinarily complex processes and biomolecular machines?

---

[40] Ellen Callaway, *Scientists Synthesize Bacteria with Smallest Genome Yet*, Nature, March 25, 2016; cited in Scientific American, https://www.scientificamerican.com/article/scientists-synthesize-bacteria-with-smallest-genome-yet/

The scriptural quote given at the very beginning of our discussion comes to mind,

"I will praise thee; for I am fearfully and wonderfully made: marvellous are thy works."[41]

[41] Psalms 139:14

3. HOW DARWINIAN EVOLUTION IS SUPPOSED TO WORK

Dawkins – one of the leading evolution proponents - defines biology as “the study of complicated [living] things that ***give the appearance of having been designed for a purpose***.”[42] [emphasis added] Darwin’s theory is described as a process of modifying living organisms through chance variations followed by natural selection which is are both unguided, blind, and purposeless trial and error processes. Darwin’s theory was based upon gross observations of various living things functioning in their respective environments. It is deceptively and seductively simple at the gross human observational level.

Darwin’s theory can be encapsulated into five principles which are: (1) there is genetic variation in the individuals of a population; (2) variations that are beneficial to those individuals in their particular environment result in an increased survival rate versus those individuals without the beneficial variation; (3) since the variations are genetic, they are inheritable by their offspring; (4) variations occur very gradually, very slowly, in continual small steps; and (5) common descent. Common descent means that there is an original ancestor of all living things and that all life-forms are descended from that ancestor. Darwin spoke of descent with modification, meaning that new species arose from their predecessor species. The newer species must, therefore, share a common ancestor. The first three principles comprise what is typically thought of as Darwinian evolution. Each of these tenets has implications.

The founding principle of evolution is that there must be variations in individuals among a larger population. We now know that these variations are a result of DNA variations at the biomolecular level. Principle #2 is better known as ‘survival of the

---

[42] Dawkins, The Blind Watchmaker, 1996, p. 1

fittest.' The evolutionary term is natural selection. Chance variations and natural selection are the presumed secret sauce of evolution. For the beneficial variation to propagate into the overall population, the variation must be inheritable in the offspring of the individual that was fortunate enough to have the chance enhancement (principle #3). This variation must then be propagated to their offspring.

Principle #4 is the source of such statements as 'given millions of years.' The assumption is that no problem is too big that thousands of small, cumulative beneficial variations over millions of years cannot solve. This principle, however, is not as simple as it seems. Lots and lots of time, countless generations, and beneficial mutations are needed. Unfortunately, mutation rates are low. As mentioned previously, the inherent error rate is around 1 per 10 B nucleotides. External causes such as chemicals or radiation may temporarily increase the error rate, but these causes are highly variable, mostly isolated, and are almost always detrimental. Evolutionists claim that the prime source of increased mutation rates is environmental stress. Unfortunately for evolutionists, "mutation rate variation has been under intense investigation for decades. Despite these efforts, little is known about the extent to which environmental stressors accelerate mutation rates." [43] Additionally, it appears that the inherent mutation rate is already optimized; lower mutation rates prevent biological diversity while higher mutation rates are detrimental.

The requirement for lots of time has been problematic. The Cambrian Explosion is a prime example. The Cambrian period began around 575M years ago. About 40M years later, the number of organisms found in the fossil record began to explode. Vast numbers of new organisms, more complex than their previous life-forms appear within 5-10M years. Every major body type found in

---

[43] Genome Research 2019, Jan., 29(1), 64-73; https://www.ncbi.nlm.nih.gov/pmc/articles/PMC6314161/

modern life had its origin in the Cambrian Explosion.[44] Ten million years is a lot of time according to most peoples' standards but it is a relative blink-of-the-eye when it comes to evolution. Some evolutionists have tried to account for this enigma through various means. The theory of punctuated equilibrium is a good example. According to this theory, life remains in stasis (no changes) until interrupted by rapid changes, followed again by stasis. Presumably, the punctuation arises from some severe environmental stress. The idea of stasis does not fit well with the idea of continual, small, cumulative changes as the source of new organisms. If continual, small cumulative changes were the correct explanation for the origin of new organisms, there should be a continual trail of intermediate life-form fossils between the organism's ancestors down to its present form. Fortuitously for evolutionists, these proposed rapid periods of change account for the lack of intermediate forms, or gaps, in the fossil record. Stephen Jay Gould, one of the proponents of punctuated equilibrium, refers to these fossil gaps as "the trade secret of paleontology."[45] In other words, to paleontologists and evolutionists, **fossil gaps present a real and continuing problem to the theory of Darwinian evolution**.

Principle #5 also has significant issues. Prokaryotes comprise the vast majority of living organisms both in the number of varieties and in the total number of cells. Prokaryotes include the domains of Bacteria and Archaea (bacteria-like, single-celled organisms) which have remained essentially the same since the beginning of life on earth. Conversely, eukaryotes generally increased in complexity over time. Complexity means that the amount of genetic information must increase proportionally to the increase in complexity. However, evolution dictates that life must adapt to its environment, not necessarily increase the amount of information

---

[44] Zimmer, Evolution, 2001, p. 324

[45] Gould, The Panda's Thumb, 1980, p. 181

it possesses or to become more complex. According to Spetner, Darwinism "pretends to account for the appearance of all the complexity in life ... [but] **no one knows of an inherent drive in living things toward greater complexity**" and "**all point mutations that have been studied on the molecular level turn out to reduce the genetic information and not to increase it**."[46] Beneficial mutations through the loss of some genetic information are the themes of Michael Behe's book, Darwin Devolves: The New Science About DNA That Challenges Evolution (2019). This will be discussed more thoroughly as we proceed. Because of these issues and others, Gould has said that "Neo-Darwinism is effectively dead."[47] However, hope for evolutionists springs eternal – Futuyma[48] reassures us that, "**just because we don't know how evolution occurred, does not justify doubt about whether it occurred**."[49] [emphasis in original] Futuyma's statement is pretty audacious – we don't know how it works, but it must work! This sounds more like a faith-based belief than a scientific theory. In summary, as Stephen Meyer puts it, ***evolution explains "only the survival of the***

---

[46] Spetner, Not By Chance!, 1998, p. 128, 138

[47] Meyer, Darwin's Doubt, 2014, p. ix quoting Gould, Is A New and General Theory of Evolution Emerging, 1980, p. 120; Stephen C. Meyer has B.S. degrees in physics and earth science. He has a Ph.D. in the History and Philosophy of Science from Cambridge University. He was a professor at Whitworth College. His books dealing with intelligent design include Signature in the Cell (2009) and Darwin's Doubt: The Explosive Origin of Animal Life and the Case for Intelligent Design (2014).

[48] Douglas Joel Futuyma is a Distinguished Professor in the Department of Ecology and Evolution at Stony Brook University in New York. He was also on the staff of the American Museum of Natural History in New York. Futuyma is the author of several textbooks. His authoritative, graduate-level textbook, Evolutionary Biology, 3rd Edition (1997) and a simplified version for undergraduates, Evolution, 4th Edition (2017) are the most noteworthy.

[49] Meyer, Darwin's Doubt, 2014, p. ix quoting Futuyma, Evolution as Fact and Theory, p. 8)

fittest [microevolution], not the arrival of the fittest [macroevolution]."[50] [emphasis added] Reluctantly, Stephen Jay Gould[51] states that **"the 'modern synthetic' view of evolution has broken down ... on both of its fundamental claims," meaning the gradual change in genes plus the exclusive reliance on natural selection leading to adaptation**. To quote Shakespeare, "something is rotten in Denmark."[52] In a few more chapters, we will analyze how Darwinian evolution really works in practice.

Gerald Kerkut, a distinguished zoologist and physiologist, made the following statement. "There are seven basic assumptions NOT mentioned during discussions of evolution. ... The first point I should like to make is that **the seven assumptions by their nature are NOT capable of experimental verification. ... The evidence that supports [the general theory of evolution] is NOT sufficiently strong to allow us to consider it as anything more than a working hypothesis**." The seven assumptions are listed here.

1. Non-living things gave rise to living material, i.e., spontaneous generation occurred.
2. Spontaneous generation occurred only once.
3. Viruses, bacteria, plants, and animals are interrelated.
4. Protozoa (single-cell) gave rise to the Metazoa (multicell).
5. Various invertebrate phyla are interrelated.
6. Invertebrates gave rise to vertebrates

---

[50] Meyer, Darwin's Doubt, 2014, p. xi

[51] Along with Dawkins, Gould is one of the most ardent evangelists for evolution. He is a paleontologist and evolutionary biologist who taught at Harvard University and worked at the American Museum of Natural History in New York. He wrote numerous books and some 300 essays in the Natural History magazine. (https://en.wikipedia.org/wiki/Stephen_Jay_Gould)

[52] William Shakespeare, Hamlet

7. Within the vertebrates, the fish gave rise to the amphibia, the amphibia to the reptiles, and the reptiles to the birds and mammals.

Keep in mind that each of the seven assumptions is just that, assumptions. There is no evidence that they actually happened or could happen. As we proceed, it will become obvious that the power of evolution is severely limited.

4. MUTATION EXAMPLES AND PROBABILITY

The first principle of Darwinian evolution is the principle of variation due to genetic mutations. Based upon our discussion of DNA, mutations are the result of a change or changes in the base sequence in DNA. Mutations can involve insertions, deletions, reversals, exchanges, etc. of the original bases. The following demonstrates the issues associated with a single base substitution and a single base insertion.

When Alexander Graham Bell first tested the telephone, he reportedly said,

"**Mr. Watson – come here – I want to see you.**"

By removing punctuation and spaces, we get the following text string.

MrWatsoncomehereIwanttoseeyou

Let us now assume that this is a sequence of DNA bases. If we read the letters 3 at a time (like a codon), we get the following genetic string.

MrW ats onc ome her eIw ant tos eey ou

A simple base substitution, an 'X' for an 'n', might look like this.

MrW ats oXc ome her eIw ant tos eey ou

It is evident that the original 'onc' has now become 'oXc'. This codon will likely code for a different amino acid. The effect of the mutation depends upon which amino acid is substituted and where it is in the protein; some amino acid changes will have little or no effect while others will be catastrophic.

A simple single-base insertion, an 'X', might look like this.

MrW ats onX com ehe reI wan tto see you

An insertion or deletion is more problematic. Notice that every codon following the insertion (or deletion) is now different. This type of error is called a frameshift, i.e., the location where each codon starts and ends has changed. Now the entire protein would be different from the point of the mutation on. As expected, this type of mutation is usually catastrophic.

Overcoming excessive improbabilities is the keystone for evolution to work. According to Dawkins, taming chance "means ***to break down the very improbable into less improbable small components arranged in a series***. No matter how improbable it is that an X could have arisen from a Y in a single step, it is ***always possible*** to conceive of a series of infinitesimally graded intermediates between them. ... And provided we ***postulate*** a ***sufficiently large series of sufficiently finely graded intermediates, we shall be able to derive anything from anything else***. ... We are allowed to do this ***only if we have sufficient time*** ... and also ***only if there is a mechanism for guiding each step in some particular direction***."[53] There it is, the entire evolutionary scheme laid out in a single, concise paragraph. Being "able to derive anything from anything else" is a ludicrous statement. "IF we have sufficient time" is always the catch-all phrase for explaining everything in evolution. "IF there is a mechanism for guiding each step is some particular direction" is the real crux of the problem. **While denying any purpose to random changes, evolutionists always talk as if there is a goal in mind!**

Several comments are in order. First, Dawkins seems to be poor at mathematics. If the probability of one thing is 1/100, and the probability of the next thing is 1/10, then the probability of both happening in a row is 1/100 X 1/10 = 1/1000. Breaking a probabilistic event into smaller sequences does NOT lower the overall probability of the total sequence occurring. To make a point, Dawkins postulated that an alien creature that lives millions of years would have a different feeling for the meaning of the chance of an event than we have. If the alien lives a hundred million years, he could have played very many hands of the card game

---

[53] Dawkins, The Blind Watchmaker, 1996, pp. 317-8

bridge. Dawkins **assumed** "it would not be unusual for him to see a 'perfect' bridge hand where each player was dealt thirteen cards of the same suit." The cards in a bridge deck are the same as in a poker deck, thirteen cards in each of four suits. "They will expect to be dealt a perfect bridge hand from time to time and will scarcely trouble to write home about it when it happens.'" Spetner states, "He is wrong. One can easily calculate the chance of Dawkins's alien experiencing a perfect bridge hand at least once in his lifetime. The chance of getting such a hand in one deal is 1 in 4.47×10^{28}. If the alien plays 100 bridge hands every day of his life for 100 million years, he will play about 3.65×10^{12} hands. The chance of his seeing a perfect hand at least once in his life is then 8.17×10^{-15} ($3.65 \times 10^{12} / 4.47 \times 10^{28}$) or about one chance in a quadrillion." [Note: $10^{-15} = 1/10^{15}$] Spetner observes that "Dawkins talked about chance, but he didn't calculate the chance of anything. Nor did he cite anyone who did. He just **assumed** that cumulative selection could lead to **macroevolution**."[54] It cannot.

Second, Dawkins admits that there should be lots of finely graded intermediates that do not occur. Third, the 'mechanism for guiding each step' is supposed to be natural selection. The only guidance is survival. Each and every infinitesimal step must be beneficial to the organism's survival in and of itself. **Guiding anything in a 'particular direction' cannot be part of evolution. Evolution cannot have any knowledge of the past or the future; there can be no goals and no purposes**. It seems to be the height of hubris to state that any process, let alone evolution, can 'derive anything from anything else.'

Let us examine some real probabilities pertaining to molecular biology. What is the probability of creating a functional protein by randomly selecting a chain of amino acids? The probability of randomly selecting the correct codons to create a specific protein occurring by pure chance is beyond astronomical. The average

---

[54] Spetner, Not By Chance!, 1998, pp. 165-6, 162-3

human protein has 375 amino acids, this, in turn, requires 375 codons. The probability of randomly creating the correct sequence of codons, which codes for 1 of 20 essential amino acids, strictly by chance is, therefore, $20^{375} = 10^{488}$. For comparison, **there are an estimated 10^{80} atoms in the entire visible universe**. If we assume a smaller protein, say 150 amino acids, the probability is **only** 1 in 20^{150} or 10^{195} of randomly generating a specific protein.

Douglas D. Axe became interested in evolutionary theory after reading Dawkins' The Blind Watchmaker. Axe concluded that "**Dawkins repeatedly smuggled in the very thing that he insisted the concept of natural selection precluded: the guiding hand of an intelligent agent**."[55] A master mathematician and protein expert, Axe[56] was interested in the probability of creating a folded, stable potentially functional protein. His area of expertise is protein folding. To narrow the search space from 20^{150} or 10^{195} for a 150 amino acid chain, he made several simplifying assumptions. The result of his calculations indicates that only one in 10^{74} random sequences is capable of folding and forming a stable protein. **Axe further estimates that only 1 in 10^{77} would be biologically useful (1 out of 1000 of the stable proteins).** To put it simply, the probability of creating a single folded, stable, function protein from a random sequence of 150 amino acids is one in 100,000,000,000,000,000,000,000,000,000,000,000,000,000,000,000, 000,000,000,000,000,000,000,000,000,000,000. The entire history of living things on the earth is dominated by bacteria both in time

---

[55] Meyer, Darwin's Doubt, p. 185

[56] Axe, *Estimating the Prevalence of Protein Sequences Adopting Functional Enzyme Folds*, Journal of Molecular Biology, Vol. 341, Issue 5, 27 Aug 2004, pp. 1295-1315; David Axe is the Maxwell Professor of Molecular Biology at Biola University. He received his Ph.D. in mathematics from Caltech before joining the Centre for Protein Engineering at Cambridge. He is the author of Undeniable: How Biology Confirms Our Intuition That Life is Designed (2016)

and numbers. Axe estimates that **there have been 10^{40} bacteria since the beginning of the world**. He generously assumes that every bacterium has 1 mutation. Therefore, he calculates that there have been 10^{40} mutations in bacteria. The probability of generating one functional protein would be $10^{40} / 10^{77} = 1/10^{37} = 10^{-37}$ in the entire history of life and all the generations of bacteria. This probability is so low it is equivalent to zero, i.e., it is impossible.

When Axe was at the Centre for Protein Engineering in Cambridge, England, he worked with a team that tried to invent new proteins, specifically enzymes. Enzymes are proteins that catalyze (enable or speed up) other chemical reactions. The scientists would intelligently replace some amino acids in the protein, then let the protein mutate. After eleven years, they wrote, "efforts to date to generate novel catalysts have primarily demonstrated that we are getting good at making bad enzymes."[57]

To quote Gelernter, "you don't turn up a useful protein merely by doodling on the back of an envelope."[58]

---

[57] Axe, Undeniable (2016), p. 42

[58] Gelernter, *Giving Up Darwin*, Claremont Review of Books, Spring 2019, p. 107; David Gelernter is not a known evolutionist or intelligent design advocate. So why does he deserve special attention? Precisely because he is not an expert in the field. However, he is an engineer with a special background. Gelernter is a professor of computer science at Yale University and a member of the National Council of the Arts. He has made major contributions to the field of parallel computing and co-founded a company that envisioned software that would aid human data management. According to The Washing Post, Gelernter is a "vehement critic of modern academia, … has condemned belligerent leftists, and blamed intellectualism for the disintegration of patriotism and traditional family values." (https://en.wikipedia.org/wiki/David_Gelernter) Something about him attracted the attention of the Unabomber who sent him a mail bomb. David's right hand and eye were permanently damaged. After reviewing Stephen Meyer's book Darwin's Doubt and others, he was prompted to write an essay entitled *Giving Up Darwin*. He

Now that we have discussed how evolution is supposed to work and how improbable it is to create even a small functional protein by chance, let us move on to discuss conscious engineering design as a prelude to the idea of Intelligent Design. This will be followed by a section discussing the details of how evolution actually works.

---

states that like so many others, "I grew up with Darwin's theory, and had always believed it to be true. ... There's no reason to doubt that Darwin successfully explained the small adjustments ... yet there are many reasons to doubt whether he can answer the hard questions and explain the big picture ... the emergence of new [species]." Gelernter feels that Meyer's arguments have demolished Darwin's theory. (p. 104) So Gelernter is one who has changed his mind, from pro-evolution to anti-evolution and looked for a replacement, be it intelligent design or something else. He does say that he cannot accept intelligent design as Meyer presents it and is sad about the loss of Darwin's beautiful ideas. This abandonment of evolution after detailed research is quite common, including several of the authors mentioned above: Behe, Meyer, Johnson, and Gelernter.

5. HOW ENGINEERING WORKS – THE DESIGN OF AN IMAGING SYSTEM

Engineering is fundamentally different than biology. Much of biology is devoted to determining how living things work. Determining the intricate details of extremely complex molecular machines is demanding, difficult, and time-consuming. However, "understanding how something works is not the same as understanding how it came to be."[59] Richard Feynman, a Noble Prize-winning quantum physicist, said "What I cannot create, I do not understand."[60] Engineering is about creating new systems through conscious processes. The premise of this book is that life came about through intelligent design. Consequently, a high-level understanding of how humans design useful systems should prove valuable, though humans are not quite as good at design as The Intelligent Designer. As opposed to the chance-driven, random, unguided, blind, and purposeless trial and error process found in evolution, engineering utilizes design, planning, goals, foresight, and purpose based upon knowledge acquired through research and development. Designed machines are obvious to humans. Upon seeing a watch, we would immediately surmise that it had been purposely designed instead of assuming that it had randomly come about by chance.

Paley's Watchmaker

In 1802, William Paley began his book Natural Theology[61] with a famous quote. "In crossing a heath, ... I found a **watch** on the ground ... that watch must have had a ***maker*** ... who formed it for [a] ***purpose***, who comprehended its construction, and ***designed*** its

[59] Behe, Darwin's Black Box, 1996. P. ix

[60] Way, Michael, *What I cannot create, I do not understand*, Journal of Cell Science, 2017, 130, pp. 2941-2

[61] William Paley, Natural Theology, 1802

use" [emphasis added]. Paley then poignantly discusses the purposeful design of the human eye. Paley's arguments about watches and eyes have been the focus of many articles and books. The cover of this book has a diagram of the inner workings of a watch. The title of two books address Paley's argument about watches directly, Dawkins' The Blind Watchmaker[62] and Broom's rebuttal, How Blind Is the Watchmaker?[63] The title of Dawkins' book seems to have been chosen to disparage the design argument. Darwin himself rather nonchalantly commented on the eye. "**How a nerve comes to be sensitive to light hardly concerns us more than *how life itself first originated*.**" "I can see no very great difficulty … in believing that natural selection has **converted the simple apparatus of an optic nerve … into an optical instrument *as perfect as is possessed by any member of the great Articulate class*** [vertebrates]."[64] [emphasis added]

Darwin assumed that many, small, cumulative variations and natural selection would have no problems turning a presumed light-sensitive spot into a human eye that is diffraction-limited (the lower resolution limit of an optical system based on the wavelength of visible light), has a wide field of view (120° radially), has enormous variable brightness sensitivity (over a range of 10^{10}, from brighter than broad daylight down to a single photon), and can distinguish ~1M colors. The eyes are part of an extraordinarily complex eye-brain image processing system that can compensate for optical issues and perform complicated image processing. It is estimated that 40%-50% of the brain is dedicated to image processing.

Darwin saw 'no very great difficulty' in creating such a complex and well-designed eye, or the brain for that matter. Darwin had a

---

[62] Dawkins, The Bind Watchmaker, 1996

[63] Broom & Neil, How Blind is the Watchmaker, 2001

[64] Darwin, Origin of the Species, 1859, pp. 157, 156; http://darwin-online.org.uk/Variorum/1859/1859-188-dns.html

lot of faith in his theory! Ironically, "many evolutionary biologists have been concerned with explaining the origin of new structures"[65] using Darwin's theory. The primary job for engineers is purposefully designing new structures that perform a useful function. Discussing how human engineers design systems should help to delineate the differences between a chance-based, trial and error 'design' process [evolution as previously described] versus a purposeful and knowledge-based design process.

For over thirty-two years, I worked as an engineer with Hewlett-Packard, Co. I was involved with teams that designed dozens of products. Most project teams include electrical engineers, mechanical engineers, computer scientists, program managers, team managers, human factors, marketing, and others. These team members were all highly trained and experienced. The design process, then, is anything but random, blind, and unguided.

The product discussed below was selected by me because of its similarity to a tiny segment of the human eye-brain system. As the lead materials engineer, I was responsible for searching out new technology, sourcing every critical part, setting up the manufacturing, improving the quality, and driving down the cost. Consequently, I saw the entire project unfold from soon after the first back-of-the-envelope concepts to the final functional product. Below is a summary.

The HP C6301C Capshare 920 Handheld Scanner

In the 1990s, Hewlett-Packard introduced a product called the HP C6301C Capshare 920 Handheld Scanner, codenamed Zorro. Before the existence of cell phones with cameras or small digital cameras, it was virtually impossible to copy papers, magazines,

---

[65] Encyclopedia of Evolutionary Biology, *Novel Structures in Animals*, 2016; https://www.sciencedirect.com/science/article/pii/B9780128000496003140

books, or the like without access to large, immobile copiers. There was no way to copy a mortgage contract at a client's home after signing, for example. The Capshare was small enough to be easily carried. It was designed to be held against the paper to be copied while holding down a button and moved in a U-like or Z-like motion over the paper to be copied. Hence, the codename Zorro. Subsequently, the image could be sent anywhere. The product was heaven-sent for mobile professionals like attorneys and sales personnel.

The product development team consisted of some twenty engineers and fifteen other team members including management, marketing, finance, purchasing, human factors, competitive analysis, etc. The key to the product was its optical imaging system and its associated image processor. HP Labs (HP's advanced research center), previously spent five years developing the basics of the imaging system and several custom image processing integrated circuits (IC's). The product development team spent another two years turning the HP Labs concepts into a manufacturable product. The first product was monochrome (black and white) and only captured images. Future products were planned to be smaller, would capture color images, and OCR text. OCR stands for optical character recognition wherein an image of text would be turned into editable text strings.

The Capshare product had two interdependent imaging systems. One system included two small area light sources and sensors. The second system was a one-pixel-wide linear sensor and its light source. See Figure 9.

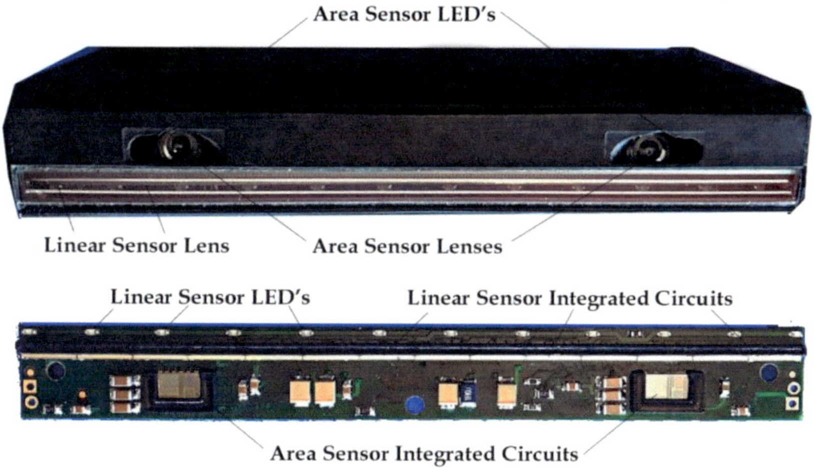

FIGURE 9. HEWLETT-PACKARD'S CAPSHARE IMAGING SYSTEM.

The top image in Figure 9 shows the bottom of the Capshare, the part of the product that makes physical contact with the paper to be scanned. From the outside of the product, one can see the area sensor lenses, linear sensor lenses, and the area sensor LED's (Light Emitting Diodes). The area and linear sensor LED's illuminate the paper. The lower image shows the internal imaging printed circuit board assembly (PCA), the area image sensors, and linear sensor LED's. There were two small area sensors and a long linear sensor.

The two area sensors were at each end of the imaging array situated about 3 inches apart and were designed to capture close-up images of the paper fibers. This was done by using high-power infrared (IR) LED's to illuminate the paper near each area sensor. The LED's were set at a high grazing angle to produce shadows of the paper fibers. It turns out that every small segment of paper has a unique pattern of fibers that can be used like fingerprints to identify any area of the paper. The linear sensor was an arrangement of 14 linear image sensors, each having 102 sensors, placed end-to-end creating a sensor 4.75 inches long. The locations of the shadows of the paper (paper fingerprints) captured by the area sensors were used to track where the 1 pixel by 1428-pixel linear image came from.

Images are often described by dpi (dots per inch). Greater dpi's make it possible to see finer image details. While printers typically print at 72 dpi, better quality images are captured at 300 dpi. At 300 dpi, each dot or pixel is ~3.3 mils (~80 um). The linear sensor had an accuracy of 300 dpi. The area sensors captured images at ~1000 dpi in a 9 X 9 array since paper fibers are much smaller than readable text.

The design goal was that a full 8.5" X 11" document could be scanned in 2-3 seconds. Since the sensor width of 4.75" was less than the width of a typical paper (8.5"-10"), the Capshare would have to scan one side of the paper and then the other side. The capture speed would then have to be ~10 ips (inches per second). At 300 dpi, that corresponds to 300 dots/in X 10 in/sec = 3000 linear images in 1 sec. The 3000 linear images, each 1 x 1428 pixels, had to be subsequently stitched together. This stitching process required custom high-speed image processing circuits.

All engineering projects entail a property that Douglas Axe calls 'functional coherence.' He defines functional coherence as "nothing more than the complete alignment of low-level functions in support of the top-level function." [66] Every low-level function in the Capshare was designed to enable it to scan an image quickly and accurately. This concept is related to Michael Behe's concept of 'irreducible complexity,' defined as a system of interacting parts that must work together to perform a function. Each part is required for the system to keep functioning, i.e., the system is irreducible – it cannot be made less complex and keep functioning.[67] The Capshare product would not work if a single part was non-functional, i.e., it was both irreducibly complex and

---

[66] Axe, Undeniable (2016), p. 143

[67] Behe, Darwin's Black Box, 2006, p. 39

functionally coherent. **All living systems are also both irreducibly complex and functionally coherent at the molecular level.**

In summary, it took more than 7 years, more than $100M, more than 40 expert people, and the most advanced technologies in the world to design a product to scan an image of a piece of paper that our eyes can accomplish almost instantaneously. Everything possible was done to design the product to specification. Nothing was left to chance or trial and error. Goals were set and plans were made to meet a specific purpose. Imagine a similar biological system coming into existence by blind chance and natural selection.

HP no longer manufactures the Capshare. However, the story has an interesting ending. Part of Capshare's technology (the area sensors) was used to design the first 'optical mouse' designed to work with computers. You probably have one connected to your computer right now.

6. HOW EVOLUTION REALLY WORKS

In Chapter 3, we outlined how Darwinian evolution is supposed to work. In summary, lots of small, genetic mutations occur. These mutations are naturally selected based upon whether or not those mutations **benefit** an organism, i.e., whether or not they made the organism **more survivable**. Given that genetic mutations are inheritable by the organism's offspring, the beneficial mutation should increase in the population of like organisms. Eventually, the mutation takes over the population, hence the organism type evolves. The unwritten, underlying assumption is that these mutations are caused by new, beneficial genetic information. We shall see that this does not occur. Instead, the new organism permanently loses genetic information or DEVOLVES!

To study various life-forms, biologists have developed a classification system or taxonomy system to describe the types of organisms. The classifications groups proceed from the most specific, smallest group to the most general, largest group. The classes of life-forms are as follows: species, genus, family, order, class, phylum, kingdom, and domain. There are three domains: bacteria, archaea, and eukarya. Bacteria and archaea are single-celled organisms. All advanced life-forms are classified in the eukarya domain. Organic evolution is supposed to enable a species to evolve into a new genus, a new family, a new order, etc. In actuality, much elegant evolutionary work studying Darwin's finches, Africa cichlids, Hawaiian fruit flies, Hawaiian honeycreeper birds, and Madagascar lemurs "demonstrates that the very best examples of evolution in nature – multiple radiations of widely diverse organisms over millions of years that each produced from ten to a thousand new species – all bog down before they reach the classification level of family". Getting bogged down means getting stuck in the mud which prevents further progress. Behe concludes "Darwin's mechanism has been wildly overrated – it is incapable of producing much biological change at all."[68] In this section, we will discuss five cases of documented evolution, or more precisely, **devolution**: Darwin's finches, polar bears, drosophila, E. Coli, and malaria.

These cases are not just interesting cases. Okay, the polar bear study is interesting. However, Darwin's finches have been held up as a superb example of Darwinian evolution for two hundred years. The fruit fly, drosophila, is the most evolutionarily studied higher organism in the world. The study cited is the most exhaustive ever undertaken about body plan genes. The E. coli study is the longest evolutionary study ever undertaken and is still ongoing. The case study of malaria demonstrates what evolution can do given trillions and trillions of generations and sets an upper limit on evolutionary power.

Darwin's Finches

While on the Galapagos Islands during his voyage on the HMS Beagle, Darwin collected birds. Among these collected birds were finches. He noticed that some of the finches had big beaks, presumably for crushing large seeds, while other birds had pointed beaks, presumably for reaching for small seeds or insects. The finches also varied in overall size. Darwin did not think much about the finches until much later. After returning to England, Darwin turned the birds over to John Gould, a famous ornithologist. Gould pronounced that many of the birds were different species of finches. Darwin formed the idea that ancestor birds had colonized the islands from South America followed by changes in the birds on each of the different, relatively isolated islands.[69]

Princeton University researchers Peter and Rosemary Grant studied Darwin's finches for 40 years (1973-2013). They studied evolutionary beak changes as a function of weather, basically

---

[68] Behe, Darwin Devolves, 2019, 171 & 155

[69] Zimmer, Evolution, 2001, pp. 21-22 & 32-3

cycles of excessive rain and drought. It is believed that some 30 or so founding finches arrived on the volcanic islands around 2M years ago. As Gould told Darwin 150 years ago, all 12 species of finches he identified were still finches. "Thus the same variation, mindless selection, and relentless evolution of birds that the Grants have recorded so admirably for decades have actually been going on for about a million generations – tens of thousands of times longer [than their study] - involving a cumulative total of billions of birds. The results of the 'net trend' produced by all that frenetic Darwinian evolution is a twofold variation in body length, shorter or longer beaks of greater or lesser depth (~5% variation), and not much else."[70] What the Grants documented was "incessant, back-and-forth selection with no real long-term changes." In many cases, one variation in a population replaces competing variations completely. In this case, two variations in beak shape live side-by-side. When the environment is wet, the finches with longer, pointy beaks out-produce those with shorter, blunt beaks. When the environment turns dry, the finches with the short beaks out-produce the long beak finches until they become the majority. This cycle repeats itself ad nauseam every time the predominant weather changes.

With the very recent developments of automated DNA sequencing and other biochemical methods, the exact ordering of DNA base pairs can now be determined and compared between organisms. Certain genes are regulatory genes; some of which switch on entire developmental pathways. The Grants determined the complete genomes of 120 Galapagos finches, close relatives, and other birds. They identified 6 master genes already known to affect head and beak development. One gene (ALX_I) was found that was the most strongly correlated to beak development in their finches. ALX_I codes a 326 amino acid protein. Two of the 326 amino acids differ between the two beak types. The pointed beak version

---

[70] Behe, Darwin Devolves, 2019, pp. 146-7

is the original; the blunt beak version is the one with the two mutations. Computer simulations of the mutations indicated that both of the mutations were **damaging to the gene, i.e., they impair the function of the normal gene**. Both versions of ALX_1 occur in the finch population. When the environment is wet, the finches with pointed beaks (unmutated) survive better and, conversely, when the environment is dry, the finches with blunt beaks (mutated) multiply faster. Consequently, **a damaging, less-functional gene serves as a beneficial mutation when the environment is dry**. The mutation is believed to have originated shortly after the South American finches arrived on the Galapagos Islands. "Apparently, ... the same two very nearly identical gene variants ... have coexisted on the islands, shifting back and forth by selection and drift, ... world without end."[71]

Darwin's finches, then, are our first example of **devolution**—not evolution. Note that in ~2M years and billions of birds, the finches are still finches to this day – they just have two kinds of beaks.

Polar Bears

The polar bear is the world's largest land carnivore. It is related to the brown bear (grizzly). The polar bear is estimated to have branched off from the brown bear 100K-200K years ago. The polar bear is adapted to cold temperatures, walking on snow and ice, swimming, and hunting seals. Seal blubber is rich in fat. In other animals, such a fat-rich diet leads to high levels of cholesterol and heart disease. It is not surprising, then, that the most strongly selected mutations had to do with fat metabolism. Multiple variations in the APOB gene – a gene involved in fat metabolism in mammals - have been detected. Since it is difficult and even dangerous to do experiments on polar bears, computer modeling

---

[71] Behe, Evolution Devolves, 2019, pp. 149-152

was used. Computer modeling indicated that the mutations were likely highly damaging.[72]

Another highly selected gene in polar bears is associated with pigmentation. LYST is the most likely gene responsible for the polar bear's white fur versus the grizzly's brown fur. Computer analysis showed that these mutations were almost certainly damaging, as well. "In fact, all of the mutations in the seventeen genes that were the most highly selected, about half were predicted to damage the function of the respective coded proteins."[73] Only 3-6 of the 17 genes were free of degrading mutations. Consequently, between 65%-85% of the beneficial, positively selected genes are estimated to have at least one damaging mutation, i.e., information is lost versus the original genome. The result is that the polar bear genome has lost information and has become more damaged versus its ancestor, the brown bear.[74] **The mutations are such that the polar bear could not become a brown bear again by simply activating different genes**. Each of the 17 genes would have to mutate in exactly the same locations, in exactly opposite ways from the original mutations – a fantastically unlikely sequence of events.

We now have grizzlies and polar bears and our second example of devolution.

Drosophila

Starting in 1979, Christiane Nusslein-Vollard and Eric Wieschaus working at the European Molecular Biology Laboratory in Heidelberg began investigating mutations in fruit flies (drosophila). They began feeding male flies a potent mutation-causing chemical. They then bred the males with female flies. The offspring larvae were examined for visible defects. The goal was to examine the effects that mutations had on regulatory embryonic

---

[72] Behe, Evolution Devolves, 2019, pp. 15-17

[73] Behe, Evolution Devolves, 2019, pp. 15-16

[74] Behe, Evolution Devolves, 2019, pp. 16-17

development, those that divide the embryo into the head, thorax, and abdomen of a fully formed adult fly.

The researchers observed the damaged larvae and inferred which specific genes regulated which parts of the fly's body plan. Their work was revolutionary because it was "the first mutagenesis in any multicellular organism that attempted to find most or all of the mutations that affect ... the essential patterning genes that are used throughout development." This study became known as the Heidelberg Screen and earned the researchers the Nobel Prize. Unfortunately, or intentionally, little was said about the fruit flies. "**Without exception, the mutants he studied perished as deformed larvae long before achieving reproductive age**." After presenting a paper at a 1982 scientific conference, Wieschaus made the following statement during the question and answer session, "The problem is, we think we've hit all the genes required to specify the body plan of Drosophila and yet these results are obviously not promising as raw materials for macroevolution." Thirty years later, evolutionists still do not know the "right mutations for major evolutionary change."[75]

Our third example of evolution shows that despite an exhaustive study of every body plan gene in drosophila, evolutionists have no idea how major body plans (macroevolution) could have taken place. In this case, the devolution of embryonic regulatory development genes was fatal in every case.

E. Coli

Evolution requires organisms, mutations, and time. The chance of getting a mutation goes up if there is more time or if there are more organisms. However, **what is most important is the number of generations**. The faster an organism reproduces, the higher the likelihood of getting a mutation. In general, larger organisms reproduce more slowly than smaller organisms. Doing

[75] Meyer, Darwin's Doubt, 2013, pp. 255-7

evolutionary experiments with elephants would require millennia, thousands of acres, and tons of food. Using bacteria, on the other hand, takes far less time, space, and food. Richard Lenski, a microbiologist from Michigan State University, has conducted probably the longest evolutionary experiment in world history using the E. Coli bacterium (February 1988 to the present). Michael Behe's description of the results of Lenski's "most definitive evolution experiment ever" is paraphrased below.[76]

The E. coli bacterium is extremely small, measuring about 0.5 um X 2 um (0.02 mil X 0.08 mil). Under favorable conditions, it can reproduce in as little as 20 minutes. Lenski decided to grow liquid cultures of E. coli in a dozen separate flasks overnight to see what would happen. Multiple cultures were chosen for replicability's sake. Overnight, the bacteria went through 6-7 generations and consumed all available food (dissolved sugar). Each flask contained some 100M bacteria.

The next day, Lenski withdrew 1% of the culture from each of the flasks and started a new set of flasks. Lenski and a plethora of grad students and postdocs would repeat this process day in and day out – for over 30 years. The number of generations has now exceeded 65K. Fortunately, Lenski let his bacteria grow without outside influence. This is important since it avoids "the bane of evolutionary studies known as investigator interference." Investigator interference occurs when "well-meaning researchers unintentionally poke and prod their study subjects in directions the investigators, but not necessarily nature, want them to go." Lenski, fortunately, took samples of his flasks every few hundred generations and froze them for later reference. Interestingly, after a few hundred generations, the descendant cells started to grow faster than the original bacteria. After 2K generations, the evolved cells grew 37% faster on average with some variations between the 12 lines. With time, the increase in the growth rate began to slow.

---

[76] Behe, Darwin Devolves, 2019, pp. 172-9

After 50K generations, the fastest line was growing ~70% faster. The important question is why the growth rates increased?

The E. coli genome is ~5.5M base pairs. Sequencing a genome of that size was impossible in the early 1990s. It was determined that the faster-growing lines were almost twice as big as the original bacteria. It was also determined that the reason they grew faster was because of delayed time between cell splitting (binary fission). Lenski demonstrated that increases in growth rates occurred in spurts rather than continuously. Some lines grew better on alternative food sources while losing the ability to metabolize the sugar ribose. One cell line developed a diminished ability to repair its DNA which turned it into a "mutator." This line had a mutation rate greater than 100X normal. Eventually, 6 of the 12 lines became mutators. In time, all the evolved cell lines lost the ability to metabolize ribose. Ribose plays many important roles in metabolism, including in the metabolic pathway that provides energy for cellular processes.

In the early 2000s, DNA sequencing became significantly cheaper and faster. Some functional genes have mobile genetic elements (MGEs) adjacent to them. MGEs are DNA sequences that can move around within the genome. The mobile element can be inserted and removed unpredictably by the cell's protein machinery which often deletes, inverts, or rearranges DNA in the process. This was presumably the cause of the loss of the ability to metabolize ribose; all 12 lines had massive deletions in the ribose gene which was adjacent to a mobile element. In 2003, after 20K generations, the team reported that 59 genes had changed their activity levels in 8 of the 12 lines. Evolutionists were ecstatic. They reported that this experiment demonstrated multiple precise beneficial mutations.

Buried in the paper was the fact that all these changes were all in the same direction and due to the alteration of a single regulatory gene (spoT). This gene governs the "stringent response," that is, spoT signals other genes that are already attuned to it that there is an onset of a starvation emergency. The stringent response signals "cells to divert resources away from growth and division and toward amino acid synthesis ... until nutrient conditions improve."[77] The resulting activity corresponds to a preexisting biochemical plan. All 8 of the lines had a single amino acid substitution with different amino acids in different locations. Since a protein is already finely tuned to a specific task, a point mutation could slightly alter its 3D conformation and reduce its effectiveness while still working for the most part. It is extremely unlikely that a point mutation would improve the activity of a protein. In summary, the growth rates increased due to point mutations in the spoT gene which governs the cell's stringent response.

Now we must answer the question, how can the loss of genetic ability be a beneficial mutation? As an example, let us take the previously mentioned loss of the ability to metabolize ribose. The ability to use multiple sources of food, such as glucose and ribose, is important to a bacterium. If glucose goes away, the bacterium can use ribose, and vice versa. The loss of the ability to use ribose may be beneficial in the presence of other food sources. Energy is saved with the spoT gene disabled. This might enable faster growth, for example – a beneficial effect. The problem is when ribose becomes the only food source! The ability to use ribose is permanently lost.

Using the latest in DNA sequencing technology in 2016, Lenski and others reported on the results of 50K generations of E. coli evolution. The entire genomes were sequenced from 2 cultures of each of the 12 lines after 500, 1000, 1500, 2000, 5000, 10K, 15K, 20K, 30K, 40K, and 50K generations – a total of 264 complete genomes.

Behe summarizes the results. "There were two tables showing genes that were the most frequently mutated. The first table listed

---

[77] Wikipedia: Stringent response; https://en.wikipedia.org/wiki/Stringent_response

15 genes that had single point mutations (one replaced amino acid) like the spoT gene discussed above. The second table showed 16 genes that suffered repeated deletions or insertions of extra DNA. These types of mutations usually kill the gene outright. The bottom line is this. After [50K] generations of the most detailed, definitive evolution experiment ever conducted, after so much improvement of the growth rate ..., after relentless mutation and selection, **it's very likely that all of the identified beneficial mutations worked by degrading or outright breaking the respective ancestor genes. And the havoc wreaked by random mutation had been frozen in place by natural selection**."[78] [emphasis added]

The mutated lines are referred to as strains, not new species. After 50K generations and billions of bacteria, the E. coli are still immediately recognizable as ... E. coli. If it looks like E. coli, and it acts like E. coli, it must be E. coli. Our fourth example is, once again, an underwhelming example of the power of Darwinian evolution. Devolution has destroyed the mutated E. coli strains abilities to utilize ribose as a food source.

The Malaria Parasite

In this chapter, we have analyzed evolution in finches, bears, fruit flies, and bacteria. In each of these cases, evolutionary change was due to the loss of genetic information which, nonetheless, was beneficial to the organism in a specific environment. This section will analyze how the malarial parasite became resistant to several anti-malarial drugs. Evolutionists cite the development of drug resistance as another superb example of the awesome power of Darwinian evolution. Malarial drug resistance was, indeed, the result of point mutations in the parasite. One point mutation conferred resistance to one drug. Two point mutations were required to confer resistance to another drug. Enough is known about these two cases to permit the calculation of the probabilities involved in both cases. Two 'beneficial' mutations in a row appear

---

[78] Behe, Darwin Devolves, 2019, pp. 172-9

to be the limit of Darwinian evolution given the limited ~4.6 B years of the earth's existence.

Most of the public is familiar with evolution through drug resistance stories such as MRSA (drug-resistant staph) and MDR-TB (drug-resistant tuberculosis). The malaria parasite is worse - it infects some 250M people resulting in 500K deaths each year. As mentioned before, the number of generations of organisms is more important than long periods of time for evolution to work. By its nature, malaria involves trillions upon trillions of generations and offers a best-case study of Darwinian evolution. This is summarized in Behe's The Edge of Evolution: The Search for the Limits of Darwinism.[79]

One of the major issues in fighting malaria is that the parasite's life cycle passes through several dissimilar steps, each one requiring a different potential treatment. It is well-known that sickle cell anemia in humans provides some resistance to the parasite. This mutation, while beneficial against malaria, creates severe accompanying health issues for humans with the sickle cell mutation. This is typical of point mutations. Based on studies of human DNA, scientists have concluded that "this particular mutation [sickle cell] has arisen independently no more than a few times in the past ten thousand years – possibly only once."[80] Our focus, however, will not be on the sickle cell trait but on mutations to the parasite itself.

Humans have conducted a losing war with malaria for centuries. In 1934, chloroquine was discovered. Humans began chemical warfare with the malaria parasite when the first widely available antimalarial drug appeared immediately after World War II. Since then, several other antimalarials have been developed. Widespread treatments of chloroquine began about 70 years ago.

---

[79] Behe, The Edge of Evolution, 2007, pp. 44-63, 135-7, 142-3

[80] Behe, The Edge of Evolution, 2007, p. 26

Each sick person has ~1T (10^{12}) malarial cells. In recent years, ~200M people (2×10^8) contract malaria each year. Consequently, there are about $10^{12} X 10^8 = 10^{20}$ malarial cells born each year. For comparison, there have been an estimated 10^{16} E. coli cells in Lenski's 30+ year experiment outlined above. It is not surprising that Wikipedia's entry on malaria states, malaria "has placed the greatest selective pressure on the human genome in recent history." [81] Conversely, with the introduction of chloroquine, humans have generated the greatest selective pressure on the malarial parasite in its history.

Unfortunately, within several decades, the nasty parasite began to become resistant to chloroquine. The question is, why? To understand why we need to understand how chloroquine works. The P. falciparum parasite feeds on hemoglobin in human red blood cells. Malaria can digest part of the hemoglobin by breaking it down into its constituent amino acids, but it cannot digest the heme part of hemoglobin. See Figure 8. Heme is not made of amino acids and is toxic to the bug. If heme accumulates in the parasite's digestive compartment, the parasite dies. Normally, malaria converts the heme to a harmless form. Chloroquine interferes with this process. For the drug to be effective, it must get into the parasite's digestive vacuole and remain there for a while. The parasite concentrates the drug in its 'stomach.' Getting the drug into the digestive vacuole is a difficult process. Consequently, there are numerous ways for the parasite to counter chloroquine's effectiveness.

Malaria's genome is ~100M base pairs that code ~5300 proteins. A mutation in one of these proteins renders chloroquine ineffective. The fateful protein, called PfCRT, was determined to be a protein pump situated in the membrane of the digestive vacuole. This protein has 424 amino acids. Mutant PfCRT proteins showed anywhere from 4 to 8 amino acid changes out of the 424. Two of

---

[81] https://en.wikipedia.org/wiki/Malaria

the changes, at locations 76 and 220, were always present and are presumed to be the changes critical for drug resistance. Thus, drug resistance to chloroquine required two mutations.

Atovaquone is another antimalarial drug that was introduced in 1996. In the same year, the first cases of resistance were reported.[82] A clinical study showed that ~1 in 10^{12} malarial cells had developed spontaneous resistance. A subsequent experiment showed the resistance was due to a single amino acid change. Thus, drug resistance to atovaquone required one mutation.

While drug resistance occurred relatively quickly to atovaquone (~1 year), resistance to chloroquine has appeared fewer than 10 times in the entire world in 50 years. That means that the average number of resistance occurrences equals once every 5 years for chloroquine.

If one multiplies the number of parasites in a single person times the number of people who get malaria each year times the number of years from drug introduction to the appearance of resistance, one can estimate the chance of a bug developing drug resistance. For chloroquine, the chance is ~1 in 10^{20} which requires 2 mutations. For atovaquone, the chance of a single mutation has been calculated to be ~1 in 10^{12}. Remember that the probability of two individual occurrences happening back-to-back is the product of the two probabilities. If we use the probability of one mutation to be ~10^{12}, as per atovaquone, then the probability of two mutations would be $10^{12} \times 10^{12} = (10^{12})^2 = 10^{24}$. If we use the probability of two mutations to be ~10^{20}, as per chloroquine, then the probability of one mutation would be $\sqrt{10^{20}} = 10^{10}$. [The √ symbol means the square root; $\sqrt{100} = 10$.]

In summary, the probability of one mutation is one in ~10^{10} to 10^{12} as with atovaquone. The probability of two mutations is one in ~10^{20} to 10^{24} as with chloroquine. The probability of two

---

82 https://www.mmv.org/malaria-medicines/history-antimalarials

beneficial mutations occurring is astronomically small, ~1/1,000,000,000,000,000,000,000,000. Behe, therefore, suggests that **two beneficial mutations are most likely the limit of Darwinian evolution** despite the trillions of generations and billions of years of life on the earth.

This does not bode well for macroevolution. The next chapter examines some selected issues that further explain the inadequacy of Neo-Darwinism to 'create anything from anything else.'

7. SELECTED ISSUES

The purpose of this section is to review five major issues. There could be many more, such as blood clotting, immunology, bat echolocation, owl's wings, the development of language, how each of the body's organs works, etc. However, these five issues should be sufficient to highlight outstanding severe issues with Darwinian evolution. The issues to be discussed are the Cambrian Explosion, a recent paper discussing a massive DNA study across 100K species, the concept of irreducible complexity using the human eye, the origin of life, and human evolution.

The issue associated with the Cambrian Explosion is the massive number of new types of life-forms that developed in an extremely short period of time by evolutionary standards. The DNA paper is a huge DNA study that shows that most of life on earth originated at the same time and not too long ago – this too discredits the premises of evolutionary theory.[83] The idea of irreducible complexity was introduced in the molecular biology chapter when discussing complicated structures such as the replisome and spliceosome. Here, the object of discussion is the human eye. The design of the eye is covered more deeply due to its importance among evolutionists who claim the eye has flaws and therefore must have developed by a trial and error process. The 'origin of life' is a key topic associated with evolution. This topic deals with the ultimate ancestor. However, if Darwinian evolution can hardly modify existing organisms, chemical 'evolution' is utterly incapable of creating life out of inanimate matter. The word 'evolution' is often attached to the word 'chemical.' This is inappropriate, however, since evolution cannot apply to non-living matter. The last issue, human evolution, is a contentious subject.

---

[83] Marlowe Hood, *Sweeping gene survey reveals new facets of evolution*, PhysOrg, May 28, 2018; https://phys.org/news/2018-05-gene-survey-reveals-facets-evolution.html

We will show that paleoanthropology is an inexact science fraught with passion, prejudice, subjectivity, and deception.

There are many details associated with each topic. Meyer's entire book, Darwin's Doubt (2014), is about the Cambrian Explosion while Behe's book, Darwin's Black Box (2006), concerns irreducible complexity. The concept of irreducible complexity means a system of complementary, interacting parts that work together to perform some function. Each part is required for the system to keep functioning, i.e., the system is irreducible – it cannot be made less complex and keep functioning. Instead of evolutionary texts, I predominantly use textbooks written by optical design engineers, vision scientists, neuroscientists, ophthalmologists, and optometrists to describe the human eye. On the origin of life, Meyer's Signature in the Cell (2009) and Thaxton's et al The Mystery of Life's Origin (2020) are consulted. An excellent resource for human evolution is Lubenow's Bones of Contention (2004).

Cambrian Explosion

The problem with the Cambrian Explosion for evolutionists is the extremely rapid development (for evolution) of organisms with major new body types. Not too many years ago, scientists thought the Cambrian Explosion took place over more than 100 MY (million years). This amount of time was more pleasing to evolutionists. As time went on, however, it has been discovered that the explosion in new life-forms took place over a mere 5-10M years. This relatively lightning-fast time-period is very problematic for evolutionary theory.

The earth is estimated by scientists to be ~4.6 B years old. Geologists and paleontologists have broken the earth's history into convenient periods of time based upon geological strata (layers). The Cambrian period started ~543 MYA (million years ago) and ended ~490 MYA. Between ~530 MYA and ~525 MYA, there was an exponential increase in organism diversification. Geologically, 5M years is ~0.1% of the earth's age. Some evolutionists estimate the elapsed time as 10M years instead of 5M. In either case, 5-10M years is a heartbeat in evolutionary time. The problem is that evolution is not supposed to be able to produce so many new and different life-forms in such a short time. Some evolutionists try to downplay the brevity of the period.

Hope springs eternal for evolutionists. Zimmer said that "the Cambrian explosion certainly happened quickly, as geological history goes, but it still took around 10 million years. And judging from the way different animal embryos are built, a few relatively small genetic changes **may** have triggered **many** of the **dramatic transformations** to animal bodies seen during that time." [84] [emphasis added] Darwin, in Origin of the Species, argued against Paley's watch-to-watchmaker analogy. To avoid the possibility of evolving entirely new and complete biological systems, i.e., whole watches at a time, Darwin proposed that evolution proceeded one tiny gear or step at a time. Consequently, he expected to see many transitional fossils in the Cambrian deposits - he did not. It was written off at the time because of an incomplete fossil record – the hope was that the missing fossils would eventually turn up. This was Darwin's 'doubt.' With so many transformations presumably going on, one would expect to see innumerable intermediate life-form fossils by now. Some 150 years after Darwin, there are still none!

While Darwin encapsulated the theory of organic evolution, many of his ideas occurred before the Origin of the Species was published. Not everyone agreed with Darwin or his predecessors. Isaac Newton, hearing these, commented, if "we men and beasts, etc. [were] made by fortuitous jumblings of the atoms there would be many parts useless in them, here a lump of flesh there a member too much. Some kinds of beasts might have but one eye, some more than two."[85] Newton's understanding of 'atoms' was

---

[84] Zimmer, Evolution, p. 324

[85] White, Isaac Newton – The Last Sorcerer, 1997, p. 149

about 100 years ahead of his time, knowing that they were tiny bodies of mass. Obviously, Newton did not know about DNA or proteins.

FIGURE 10. DARWIN'S PROPOSED TREE OF LIFE VERSUS THE ACTUAL TREE OF LIFE FOR THE CAMBRIAN EXPLOSION. [86]

Figure 10 illustrates Darwin's proposed tree of life (left diagram) and the tree of life evidenced by the Cambrian fossil record (right diagram). Darwin's proposed tree of life was a branching structure, with new types of organisms branching off as time went on. Unfortunately for proponents of evolution, the tree of life as evidenced by Cambrian fossils is instead a series of very slightly changing life-forms. Figure 11 depicts a display at the California Academy of Science in San Francisco, California. I took this photograph back in the 1990s. The display was designed to demonstrate Darwin's proposed tree of life. Notice that at each branch or intersection (the Y's), there is supposed to be a transitional life-form. Notice that there is a magnifying glass at each intersection. Look closely, there are NO fossils under any of the magnifying glasses – guess they are still looking for them. Once again, there are no transitional forms.

---

[86] Modified by D. John Redman. By James St. John - Morrison Formation (Upper Jurassic; No Thoroughfare Canyon, Colorado National Monument, Colorado, USA) 3, CC BY 2.0, https://commons.wikimedia.org/w/index.php?curid=48175889

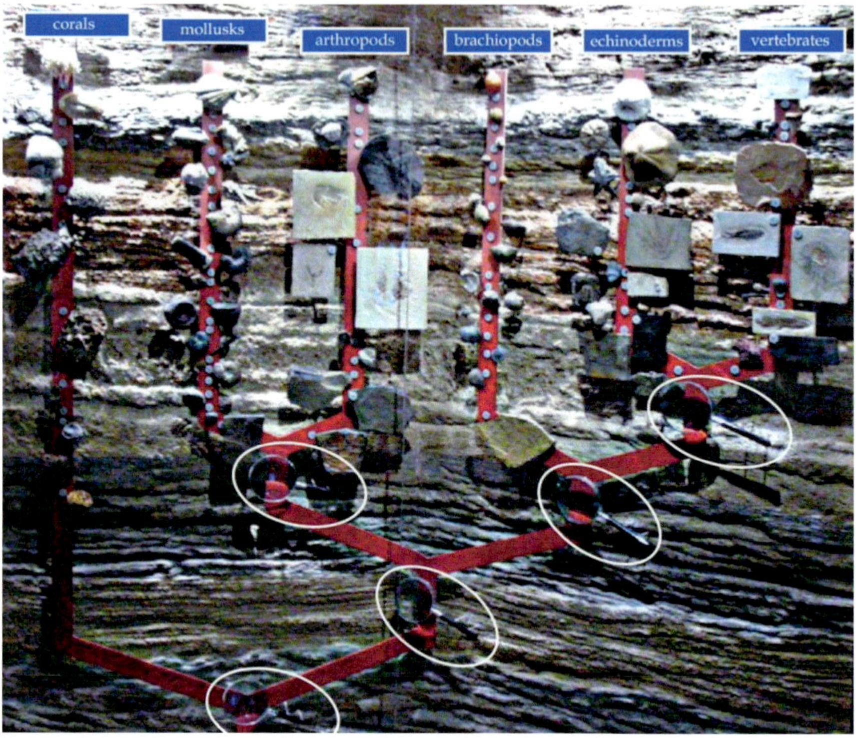

FIGURE 11. A PHOTOGRAPH OF A PRESUMED EVOLUTIONARY TREE OF LIFE AT THE CALIFORNIA ACADEMY OF SCIENCES. [87]

There are three major evolutionary problems illuminated by the Cambrian Explosion: (1) the development of numerous new body plans, not just simple adaptations, i.e., macroevolution; (2) the rapid rate of development; and (3) the absence of transitional fossils. **Notably, the Cambrian Explosion is not an isolated instance; it is just the biggest and most well-known**.

The theory of evolution implicitly and explicitly incorporates the idea of a common ancestor. Evolutionists explain that the evidence for this includes statements like the following. "All currently living organisms on Earth share a common genetic heritage ... 6,331 groups of genes common to all living animals have been identified; these may have arisen from a single common

[87] Photograph by D. John Redman.

ancestor that lived 650M years ago in the Precambrian."[88] The biological artist, Goodsell states that a single bacterial cell "uses much the same machinery as your own cells." He states further that there is a "common birthright of molecular machines."[89] These claims center around common biochemistry, common genetic code, the fact that all organisms use the chemical compound ATP (adenosine triphosphate) as an energy carrier, and that all amino acids in proteins are left-handed. Some of these may have come about because of the laws of physics and chemistry as opposed to biological evolution or because the same environmental conditions caused the same biochemistry to evolve convergently (convergent evolution).

In graduate school, I was taught to always look for possible alternate explanations for a hypothesis! There is an alternate explanation for a common ancestor – it is called 'reusable code.' Computer programmers write software code for all kinds of digital hardware systems (computers, appliance controllers, smartphones, etc.). Repeatedly, they must control the same or similar hardware. For efficiency reasons, programmers will have libraries of code to use plus much of their personally developed code. It seems more than reasonable that an intelligent designer would do the same. When the intelligent designer has developed an optimized energy delivery system, e.g., ATP, why not reuse it? **Everything that appears to have arisen from a common ancestor can be alternately explained as reusable DNA code**.

9 Out of 10 Species on Earth Today Came About at the Same Time and Species are Genetically Isolated

If one is watchful, you can see all sorts of newspaper and magazine articles that talk about evolution – presumably to

---

[88] https://en.wikipedia.org/wiki/Common_descent

[89] Goodsell, The Machinery of Life, pp. 2-3

popularize the theory. If one reads more carefully, however, some articles tell a deeper story. Sometimes one wonders why they even published it since the content is so unfavorable towards evolutionary theory. The following is such an article. Phys.Org[90] is an online magazine that publishes articles concerning all the sciences. On May 28, 2018, the following article appeared: *Sweeping Gene Survey Reveals New Facets of Evolution*. [91] Unfortunately, the new facets do not jibe well with traditional Darwinian evolution. The full article was published in the journal Human Evolution (May 2018).[92]

Two researchers at Rockefeller University, Mark Stoeckle, and David Thaler, **analyzed 5M gene snapshots called 'barcodes' from 100K animal species** derived from the work of hundreds of scientists around the world. This corresponds to an average of 50 snapshots per species. This data is stored in the US government-run GenBank database.

The article's author states that the results are "**sure to *JOSTLE*, if not *OVERTURN*, more than one *settled idea* about how evolution unfolds**." [emphasis added] Hmm, jostle and overturn presumably already settled ideas (treated as facts) about how evolution is supposed to work! "Textbook biology [describes how] species with large, far-flung populations – think ants, rats, humans – will become more genetically diverse with time. But is that true? The answer is no. ... for the planet's 7.6 billion people, 500 million house sparrows, or 100,000 sandpipers, **genetic diversity 'is about the same.'**" Now, here is the really good part. "The study's most

---

[90] https://phys.org/

[91] Marlowe Hood, *Sweeping gene survey reveals new facets of evolution*, PhysOrg, May 28, 2018; https://phys.org/news/2018-05-gene-survey-reveals-facets-evolution.html

[92] Stoeckle, M. Y., and Thaler, D. S., *Why Should Mitochondria Define Species?*, Human Evolution, Vol. 33, No. 1-2 (2018)

startling result, perhaps, is that ***9 out of 10 species on Earth today, including humans, came into being 100,000 to 200,000 years ago.***" Wow! "How does one explain the fact that 90% of animal life, genetically speaking, is roughly the same age?"

Genomes are large – the human genome has ~3 B base pairs grouped into ~20K genes inside the nucleus of every cell. However, all animal cells also have DNA in their mitochondria. Mitochondria are encapsulated molecular machines inside the cell that convert food into ATP for energy use throughout the cell. Mitochondria only have 37 genes, much, much smaller than nuclear genomes. One of these 37 genes is referred to as COI. The COI gene is the one used for DNA **barcoding**. There are two good things about mitochondrial DNA used for evolutionary studies: (1) it is a lot smaller, simpler, and cheaper to isolate than nuclear DNA; and (2) all animals have the same basic set of mitochondrial DNA. Though similar across all animals, the mitochondrial DNA from each species is distinctly different and can be compared to each other. Evolutionists think they know the fundamental mutation rate in mitochondrial DNA, consequently, they can determine how old the DNA is.

The article concludes with the following: "Another unexpected finding from the study [was that] **species have very clear genetic boundaries**, and there's **nothing much in between**. If individuals are stars, then species are galaxies. They are compact clusters in the vastness of empty sequence space. **The absence of 'in-between' species is something that also perplexed Darwin**." [emphasis added]

There you have it, everything we have been discussing is brought into sharp relief! No branching tree of life, most modern life-forms came about at the same time (long after the Cambrian Explosion), reusable code (mitochondrial DNA), and no transitional or 'in-between' species. In summary, **Neo-Darwinian evolution does not work as advertised – there is very limited genetic diversity, 90% of life arose at the same time versus gradually, and**

there are clear genetic boundaries within species, with virtually no life-forms in between (no intermediates).

Irreducible Complexity

Irreducible complexity has been introduced previously. This section will discuss the human eye for two reasons. The first reason is that the human eye is a perfect example of an irreducibly complex biological system. The second reason is that, according to evolutionists, there are presumed design flaws in the eye that supposedly prove that the eye came about through evolution versus an intelligent designer. The assumption is that an intelligent designer would not make such silly mistakes, however, these kinds of mistakes are part and parcel with evolution since it operates as a blind, trial and error process. We shall see that every presumed design flaw is, in reality, a brilliant design feature. These design features demonstrate that evolution could not have generated the eye – it took an intelligent designer.

Michael J. Behe is the creator of the idea of irreducible complexity which is the principal subject of his first book, Darwin's Black Box (2006). He defines an irreducibly complex system as follows: "by irreducibly complex I mean a system composed of several well-matched interacting parts that contribute to the basic function, wherein the removal of any one of the parts causes the system to effectively cease functioning." Continuing, "an irreducibly complex system cannot be produced directly by slight, successive modifications of a precursor system."[93] In other words, the system cannot be continuously improved while continuing to work by the same mechanism. This means that **Darwinian evolution cannot create irreducibly complex systems**. This is for several reasons. One is that evolution is random; there cannot be any foresight into what is needed. It cannot know that a second, or third, or tenth part is needed to make the system work. Second, given the extremely low probabilities of beneficial mutations, the

---

[93] Behe, Darwin's Black Box, 2006, p. 39

probabilities of simultaneously getting multiple beneficial mutations are essentially zero. A third reason, the coup de gras, is that evolution must be beneficial at every step to be naturally selected. This means that at every single mini step, every part must continue to interact in such a way as to make the entire system beneficially functional.

Dr. Behe uses a simple mousetrap as a practical example of irreducible complexity. The common mousetrap has a minimally acceptable set of parts: a bar that is used to trap the mouse, a spring that propels the trapping bar, a piece of metal that retains the trapping bar in a spring-loaded position, and a trigger which attaches to the metal retainer. When a mouse activates the trigger, usually by tugging at food attached to the trigger, the metal retainer is released. This, in turn, releases the trapping bar. The force of the spring firmly traps the mouse. Now, imagine how the mousetrap would work without a trigger? The spring? The trapping bar? Now, further imagine, how the trap could function as it was being built one part at a time?

The Human Eye

This section will summarize the evolutionist versus creationist or intelligent design debate that has been going on for over 150 years. The evolutionist's argument is "a 'gotcha' argument: no designer would [have made these design mistakes], so Darwinism is true; so there." [94] We will proceed by summarizing the evolutionists' views on the human eye and then move to more reliable optical and vision textbooks versus evolution or biology textbooks. These textbooks include Modern Optical Engineering (Smith, 2000), Visual Perception (Schwarz, 1999), Vision Science (Palmer, 1999), Foundations of Vision (Wandell, 1995), and The Human Eye (Oyster, 1999). Because two different faiths and two

---

[94] Behe, Darwin's Black Box, 2006, p. 49

different worldviews are at stake – the vertebrate eye is the quintessential example of these dichotomies.

Since I use optical designers, vision scientists, optometrists, ophthalmologists, and neurobiologists to make my case instead of evolutionary biologists, short introductions of these experts should prove enlightening. My Master's Thesis at the Rochester Institute of Technology in Imaging and Color Science concerned visual perception. I utilized the texts below as some of my references.

1. Warren J. Smith was president of the Optical Society of America in 1980. Warren Smith is world-renowned in the optics community as the author of two editions of Modern Optical Engineering - The Design of Optical Systems[95] (2000) and Modern Lens Design - A Resource Manual and Practical Optical Systems Layout (2004). He won an Academy Award (an Oscar) for his lens design that corrected an optical error in movie projection lenses. I have taken an optics short course with Warren. Therefore, I have personal knowledge of his expertise in his field.
2. Steven H. Schwartz, OD, Ph.D., is a professor of biological and vision sciences, SUNY, College of Optometry. He is a practicing optometrist and has a passion for the development of textbooks on visual perception and optics. His books are Geometrical and Visual Optics: A Clinical Introduction, 2nd Ed. (2013) and Visual Perception: A Clinical Orientation, 5th Ed.[96] (2017).
3. Stephen E. Palmer is a Professor of Psychology and Director of the Institute of Cognitive Studies at the University of California, Berkeley. His book Vision Science – Photons to

[95] Smith, Modern Optical Engineering, 2000

[96] Schwartz, Visual Perception: A Clinical Orientation, 5th Ed., 2017

Phenomenology[97] (1999) is a reference work on the visual system.

4. Brian A. Wandell joined the Stanford Psychology faculty in 1979 and is a member, by courtesy, of Electrical Engineering, Ophthalmology, and the Graduate School of Education. He is Director of Stanford's Center for Cognitive and Neurobiological Imaging and Deputy Director of the Wu Tsai Neuroscience Institute. His book Foundations of Vision[98] (1995), is a classic on how the eye-brain visual system works.

5. Clyde W. Oyster's book The Human Eye: Structure and Function[99] (1999) is used as an optometry textbook emphasizing anatomy and physiology as well as brain function. He is Professor Emeritus in the Department of Physiological Optics of the School of Optometry at the University of Alabama at Birmingham. The university lists him as a pioneering retinal researcher.

These five scientists are eminently qualified to comment on the design and function of the human eye. They are far more qualified than any evolutionary biologist on this subject.

The vertebrate eye is a marvelous optical system. See the diagram of the eye and the organization of the retina in Figures 12 and 13. Newton opined, "Was the eye contrived without skill in optics?"[100] For our discussion, the most important part of the eye is the retina. According to Schwartz, the retina is an "exceedingly complex, multilayered neural element." Its primary purpose is to capture the image formed by the preceding optical elements of the

---

[97] Palmer, Vision Science – Photons to Phenomenology, 1999

[98] Wandell, Foundations of Vision, 1995

[99] Oyster, The Human Eye: Structure and Function, 1999

[100] Newton, Opticks, pp. 369-370

eye (the cornea, an aqueous layer, the lens, and the vitreous layer). Its secondary purpose to perform "the complex task of analyzing these images." [101] The eye, however, does not operate alone. **"Retinal neurons develop from the same cells that give rise to the brain; hence, the retina is part of the central nervous system."** [102] **It is estimated that between 40%-50% of the brain is dedicated to visual processing.**[103] For our purposes, we will not be discussing the details of the visual cortex. Instead, we will focus on the design of the retina itself since that is where the evolutionists find fault. It should be apparent, however, if evolution cannot fashion the eye, it certainly cannot fabricate the associated neural processing center.

In Evolution: The Triumph of an Idea, Zimmer begins his discussion of the eye with the caveat, "**because evolution can only tinker, it cannot produce the best of all possible designs**." Continuing, he says, "our eyes, for example, are certainly impressive video cameras, and yet in some ways they are **fundamentally flawed**." [emphasis added] Zimmer then rattles off the details that the photoreceptors are pointed backwards and the optic nerve's attachment to the retina causes a blind spot.[104] In the first place, evolution cannot tinker; tinker implies conscious effort even if the tinkerer is unskilled - evolution is completely random. In the second place, evolutionists emphatically state that the eye is flawed. We will see what optical engineers, directors of cognitive and neurobiological imaging, ophthalmologists, optometrists, and vision scientists have to say. But first, here are some of the evolutionists' comments about the presumed faulty design of the retina.

---

[101] Schwartz, p. 4

[102] Wandell, p. 111

[103] Palmer, p. 24

[104] Zimmer, Evolution: The Triumph of an Idea, 2001, p. 128

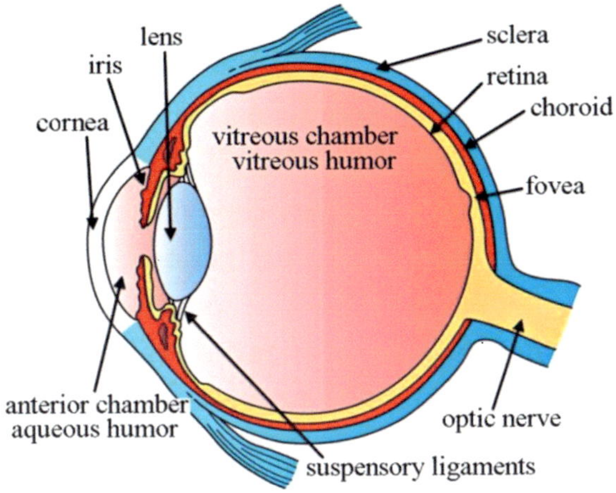

FIGURE 12. A DIAGRAM OF THE COMPONENTS OF THE EYE. [105]

Charles Darwin studied Pastor William Paley's book, Natural Theology (1802), in college. Very early in his theology, Paley states that "besides that conformity to optical principles which [the eye's] internal constitution displays, and which alone amounts to a manifestation of intelligence having been exerted in the structure; besides this, ... there is to be seen, in every thing belonging to it and about it, an extraordinary degree of care, an anxiety for its preservation, due, if we may so speak, to its value and its tenderness."[106]

In Darwin's On the Origin of the Species (1859), there is a section entitled "Organs of Extreme Perfection and Complication"[107] "To **suppose** that the eye with all its inimitable contrivances...could have been formed by natural selection, **seems**, I freely confess, is **absurd in the highest degree**." Continuing, "yet reason tells me, that **if** numerous gradations from a perfect and

---

[105] By Artwork by Holly Fischer - http://open.umich.edu/education/med/resources/second-look-series/materials - Eye Slide 3, CC BY 3.0, https://commons.wikimedia.org/w/index.php?curid=24367145

[106] Paley, Natural Theology, 1802, p. 23

[107] Darwin, On the Origin of the Species, 1859, pp. 186-7

complex eye to one very imperfect and simple, each grade being useful to its possessor, can be shown to exist; **if** further, the eye does vary ever so slightly, and the variations be inherited, which is certainly the case; and **if** any variation or modification in the organ be ever useful to an animal under changing conditions of life, then the difficulty of **believing** that a perfect and complex eye could be formed by natural selection, though insuperable by our **imagination**, can hardly be considered real." [emphasis added] A 'just-so' story is an unverifiable narrative explanation of something. Darwin has just stated a 'just-so' story – 'suppose,' 'seems,' 'absurd,' 'if,' 'if,'' 'if,' 'believing,' 'imagination.' According to Behe, **Darwin's comments were "a flight of pure imagination," not science.**[108]

Evolutionists' primary point of contention is the design of the retina. A secondary contention is a blind spot in the eye where the optic nerve leaves the eye for the brain. Dawkins thinks that an "engineer would naturally assume that the photocells would point towards the light, with their wires leading backwards towards the brain. He **would laugh at any suggestion that the photocells might point away from the light**, with their wires departing on the side nearest the light. Yet this is exactly what happens in all vertebrate retinas. **Each photocell is, in effect, wired in backwards**, with its wire sticking out on the side nearest the light. The wire has to travel over the surface of the retina, to a point where it dives through a hole in the retina (the so-called "blind spot") to join the optic nerve."[109] [emphasis added]

Similarly, G. C. Williams claimed **the retina is not just an example, but one of the best examples, of "poor design" in vertebrates** that ***proves*** the **blind watchmaker created life** and not an intelligent creator: "**Every organism shows features that are functionally arbitrary or even maladaptive**. . . . My chosen classic

---

[108] Behe, Darwin Devolves, xxx, p. 48

[109] Dawkins, The Blind Watchmaker, 1986, p. 93

is the vertebrate eye. It was used by Paley as a particularly forceful part of his theological argument from design. As he claimed, **the eye is surely a superbly fashioned optical instrument**. It is also something else; **a superb example of maladaptive historical legacy**. ... Unfortunately for Paley's argument, **the retina is upside down**. The rods and cones are the bottom layer, and light reaches them only after passing through the nerves and blood vessels."[110] [emphasis added]

Tufts University professor Daniel Dennett argued that, although the design of the eye is brilliant, "it betrays its origin with a tell-tale flaw: **the retina is inside out**. The nerve fibers that carry the signals from the eye's rods and cones (which sense light and color) lie on top of them, and have to plunge through a large hole in the retina to get to the brain, creating the blind spot. **No intelligent designer would put such a clumsy arrangement in a camcorder**, and this is **just one of hundreds of accidents frozen in evolutionary history** that ***confirm*** the **mindlessness of the historical process**."[111] [emphasis added]

Now that we have heard the arguments from the evolutionists' point of view, let us examine what the optical and vision scientists have to say. We will work our way from the front of the eye to the back of the retina for two reasons: (1) to give you some feel for the complexity and the superb imaging capabilities of the eye; and (2) to answer the evolutionists' accusations noted above. To begin, Palmer states that' "**each of the eye's components performs a critical role in the eye's sensory capabilities**."[112] [emphasis added]

---

[110] G. C. Williams, Natural selection: Domains, levels, and challenges, 1992, p. 72

[111] Dennett, Show Me The Science, The New York Times, August 28, 2005; https://www.nytimes.com/2005/08/28/opinion/show-me-the-science.html

[112] Palmer, p. 27

Light enters the eye through the cornea, it then passes through an aqueous layer, followed by the lens, and finally the vitreous layer before striking the retina. Optically, the cornea and the lens are the most significant. The lens itself is adjustable in shape, muscles attached to the edges of the lens pull or relax the lens, either thinning or fattening the lens. Consequently, the eye can typically focus on objects from infinity down to ~3 inches (7 cm) away. By comparison, a typical digital SLR camera lens can focus from infinity down to about 13 inches (33 cm). To focus closer than that requires a lens change. Most human fabricated lenses have spherical surfaces. That is because they are easier and less expensive to make. No off-the-shelf lens can change shape like the eye's lens does – they are rigid, composed of glass or plastic. The index of refraction is a number that represents how much a material can bend light. Most lenses are homogenous which means the material is the same throughout. This, in turn, means that the index of refraction is the same everywhere in the lens. The human lens is a gradient index lens wherein the index of refraction is higher in the center of the lens. These anomalies have a purpose. Smith states that "an often-overlooked fact is that the commonly accepted eye data ... **does not give an adequate picture of the *quality* of the visual system**. First, the surfaces of the eye are not spherical ... in general, the surface tends to be weaker [less refracting] toward the margin of the surface. Second, the index of the lens is not uniform but is higher in the central part of the lens. ... Note that both the gradient index and the surface asphericities introduce overcorrected spherical aberration, which **offsets** the undercorrected spherical [aberration] of the outer surface of the cornea [which is also aspherical]." [113] [emphasis added] An aberration is an imaging error. Spherical aberration is an error in the focal point, e.g., the center of the lens focuses the image at a

---

[113] Smith, p. 127

slightly different place than the edges do. How did evolution know about these errors let alone how to correct them'?

The eye has a phenomenal ability to adapt to different light levels, from less than starlight to more than full daylight. To adjust the brightness setting of a camera, a photographer changes the f-stop. The f-stop changes the size of the aperture or opening through which the light enters the camera. Every doubling/halving of an f-stop allows half/twice as much light to enter the camera, e.g., f/1 to f/2 or f/4 to f/8. Lower f-stops let in more light than higher f-stops. The eye does this by adjusting the diameter of the iris, hence, the pupil is analogous to a camera's aperture (f-stop). A typical camera can change its exposure by ~4000X with f-stops alone. Comparatively, the iris can change the exposure by only ~20X. However, the eye has two other tricks up its sleeve. First, the eye has two overlapping imaging systems, a low-light system (using rods, gray-level photoreceptors) and a bright-light system (using cones, color photoreceptors). This system can compensate for a $\sim 10^3$ change in light level. Second, the photoreceptors can decrease their sensitivity to light as the brightness increases in a logarithmic fashion.[114] This light 'adaptation' can further accommodate light levels over a range of $\sim 10^6$. These three factors permit the eye to operate over an enormous brightness range of $2 \times 10^1 \times 10^3 \times 10^6 = \sim 10^{10}$, a range of ~10 B.[115] In more familiar terms, the eye can accommodate a variation in brightness of approximately 10,000,000,000X!

When light reaches the back of the eye, it impinges on the retina and the complexity of vision gets even more interesting. Light traverses the optical nerves and blood vessels, ganglion cells, amacrine cells, Müller cells, bipolar cells, and horizontal cells before illuminating the photoreceptors. See Figure 13. Beyond the

---

[114] Palmer, p. 31

[115] Oyster, p. 529

photoreceptors is the opaque pigmented epithelium. A general overview of the structure of the retina is in order.

The high-resolution, color imaging center of the retina is called the fovea. The fovea only has cones, about 8M of them (cones are spaced very closely for high-resolution imaging and can capture color and contrast information). As we move away from the fovea, the number of cones decreases, and their spacing increases. As we move farther out, the retina is dominated by rods, about 120M of them (rods capture contrast information from dim light).[116] There are ~1M nerve fibers culminating in the optic nerve, consequently, there are more photoreceptors than nerves (~128:1). This means that multiple receptors are attached to a single nerve fiber except in the center of the fovea where each cone has its own dedicated nerve fiber.[117] Let's continue to address the evolutionists' poor design arguments about the photoreceptors presumably being wired backwards and the light having to traverse the top layer of nerves and blood vessels. Schwartz states that "**the inner retinal elements [the nerve stack] do little to degrade the retinal image because they are somewhat transparent and ... laterally displaced [away from] the fovea**."[118] [emphasis added] What this means is that the normal retinal covering is thinned over the fovea by ~1/2. Continuing, Schwartz explains, "these adaptations – absence of rods, maximal density of cones, pushing aside of inner retinal elements to expose the cone outer segments to light, and absence of retinal vasculature - ***all contribute to the excellent resolution provided by the fovea***."[119] [emphasis added] Note that there are no semi-opaque blood vessels on the top of the retina in the foveal area.

---

[116] Palmer, p. 128

[117] Schwartz, p. 128

[118] Schwartz, p. 8

[119] Schwartz, pp. 14-16

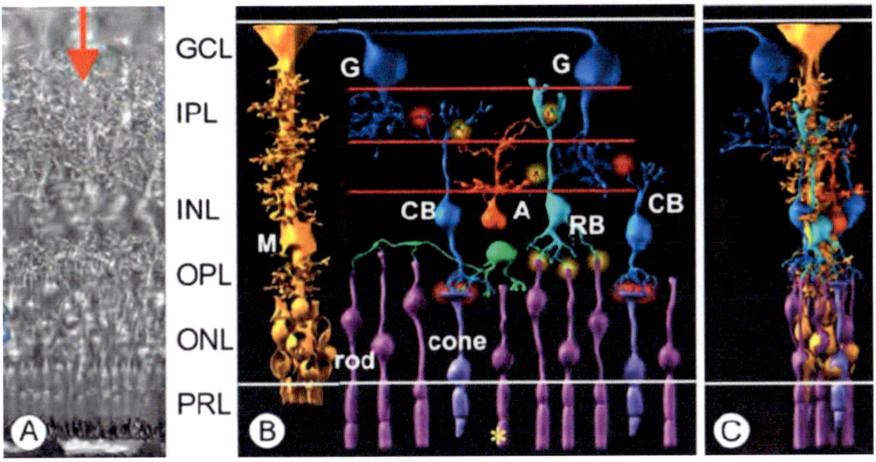

FIGURE 13. RETINAL STRUCTURE AND CELLULAR ORGANIZATION. [120]

Figure 13 is an artist's conception of the typical vertebrate retina. Panel (A) is a cross-section through the mammalian retina. The red arrow at the top-left indicates the direction of the incoming light. The light must pass through several layers: the GCL = ganglion cell layer; IPL = inner plexiform layer; INL = inner nuclear layer; OPL = outer plexiform layer; ONL = outer nuclear layer; and the PRL = photoreceptor layer. Panel (B) shows the types of neurons found in the vertebrate retina. The Müller cell (M) extends through the entire retina, top to bottom, and has radial branches that support adjacent neurons. The other neuron types include: G = retinal ganglion cells; RB = rod bipolar cells; CB= cone bipolar cells; and A = amacrine cells. In Panel (B), they are spread out for easier viewing. Panel (C) shows how the neuron types depicted in (B) cluster around the central Müller cell forming a columnar structure. These layers, however, do not impede incoming light significantly.

---

[120] Reichenbach, Andreas & Franze, Kristian & Agte, Silke & Junek, Stephan & Grosche, Antje & Grosche, Jens & Savvinov, Alexey & Guck, Jochen & Skatchkov, Serguei, (2012), *Live Cells as Optical Fibers in the Vertebrate Retina*, Selected Topics on Optical Fiber Technology, DOI 10.5772/26945; permission granted, Kristian Franze.

"It can be summarized that **the retina is all but perfectly transparent**."[121]

Palmer details what he refers to as three "peculiarities of retinal design." The first is that the "receptor cells are actually the last [layer in the retina]. The second peculiarity is that not only are the receptors last, both rods and cones appear to be pointing backward." "Luckily for our visual abilities, **the retinal cells and axons in front of the receptors are fairly transparent**, so the optical quality of the image does not suffer as much as one might imagine." [122] [emphasis added] Also, we will see in the next paragraph that these first two peculiarities are overcome by a recently discovered optical waveguide system in the retina. As the receptors are activated by incoming photons, the cones change from a deep purple color that absorbs light to a transparent state where light passes through the cone. The absorption of light also causes the pigments to change shape. Enzymes are needed for pigment regeneration (to change their shape back to active mode). These are found in the pigment epithelium which is opaque and at the very back of the retina. The third peculiarity is the presence of blind spots in both eyes as the optic nerves leave the eye. "There are two reasons for this: First, the blind spots are positioned so that receptors in one eye register what is missed in the blind spot of the other eye. Second, the visual system fills in the region with appropriate sensory qualities, just as it does the shadows of the blood vessels."[123]

---

[121] Reichenbach, Andreas & Franze, Kristian & Agte, Silke & Junek, Stephan & Grosche, Antje & Grosche, Jens & Savvinov, Alexey & Guck, Jochen & Skatchkov, Serguei, (2012), *Live Cells as Optical Fibers in the Vertebrate Retina*, Selected Topics on Optical Fiber Technology, DOI 10.5772/26945, p. 251.

[122] Palmer, pp. 32-33

[123] Palmer, p. 34

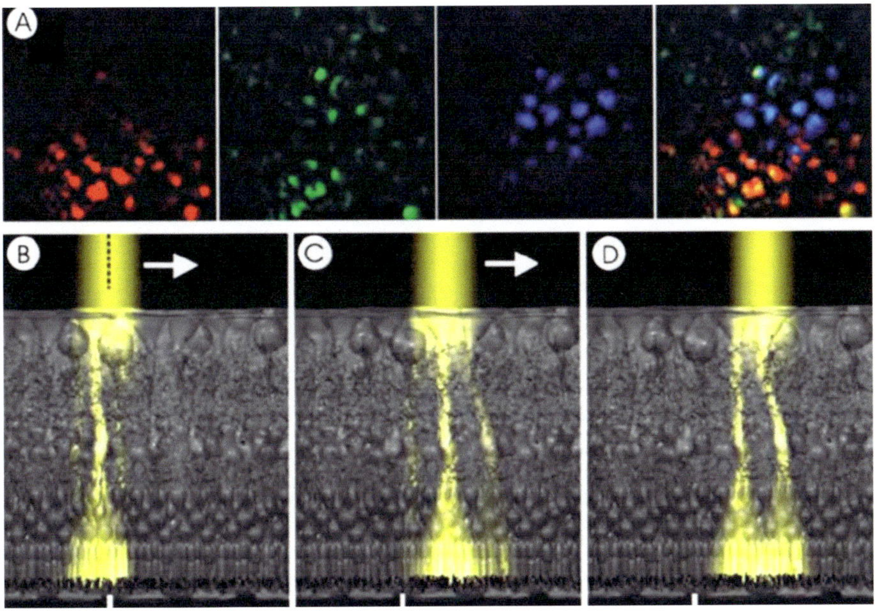

FIGURE 14. IMAGE OF THE WAVEGUIDE EFFECT IN THE RETINA. [124]

A recent discovery is of interest concerning the photoreceptors being at the back of the retina. A waveguide or optical fiber is a system that conveys light from one end of the fiber to the other end. The light spectrum refers to the different wavelengths of light which correspond to different colors of light, aka the rainbow. See Figure 14. "It has been suggested that **Müller cells function as optical fibres in the retina, transferring light illuminating the retinal surface onto the cone photoreceptors**. Here we show that Müller cells are **wavelength-dependent waveguides**, concentrating the green-red wavelengths [the most important colors for color vision] of the visible spectrum onto cones and allowing the blue-purple part [the most important colors for gray-level or contrast vision] to leak onto nearby rods. ... light

---

[124] Reichenbach, Andreas & Franze, Kristian & Agte, Silke & Junek, Stephan & Grosche, Antje & Grosche, Jens & Savvinov, Alexey & Guck, Jochen & Skatchkov, Serguei, (2012), *Live Cells as Optical Fibers in the Vertebrate Retina*, Selected Topics on Optical Fiber Technology, DOI 10.5772/26945; permission granted, Kristian Franze

propagation by Müller cells through the retina can be considered as an integral part of the first step in the visual process, increasing photon absorption by cones while minimally affecting rod-mediated vision."[125] The columnar structures depicted in Figure 13 (C) form the light guides. Figure 14 (A) shows various colors of light illuminating the retina. Panel (B) shows the red-green frequencies being guided to the center cone. Panel (C) depicts the addition of blue frequencies. Part of the light is being guided further away from the centerline to where the rods are. Panel (D) shows the full spectrum of light being separated and guided to the cones and the rods. So much for being a design flaw. Instead, **the retina efficiently separates the frequencies of light for maximum absorption and pipes the light directly from the surface of the retina to the appropriate photoreceptors at the back of the retina!**

To complete our discussion of the eye, a few parting comments should be helpful regarding how image information is passed from the retina to the central nervous system. As a review, light enters the top of the retina and chemically activates the cones and rods. These photoreceptors then convert these chemical reactions into electrical signals which pass back up the retinal neurons and over to the optic nerve. From there, the electrical signals are passed to the visual cortex where the brain continues to process them. One of the retina's tasks is to preprocess visual information before sending it to the brain. The interconnections between the different types of retinal neurons, where the preprocessing occurs, is

---

[125] Evolution News & Science Today reporting on Phys.org article, Specialized Retinal Cells Are a "Design Feature," Showing that the Argument for Suboptimal Design of the Eye "Is Folly," Casey Luskin, August 8, 2014; https://evolutionnews.org/2014/08/physorg_special/ ; original paper: Amichai M. Labin, Shadi K. Safuri, Erez N. Ribak, and Ido Perlman, "Müller cells separate between wavelengths to improve day vision with minimal effect upon night vision," Nature Communications, DOI: 10.1038/ncomms5319. July 8, 2014)

extraordinarily complex and not fully understood. The entire process is miraculous.

Palmer exclaims, "How photoreceptors manage to change the electromagnetic energy of photons into neural activity ... is a **complex and truly ingenious process**." [126] [emphasis added] Wandell explains, "evidently one of the important functions of the retina is to **organize the information encoded by the photoreceptors into a collection of visual streams**. Presumably, the purpose of these separate streams is to communicate relevant information efficiently to brain areas engaged in specialized types of visual processing." It is estimated that there "**may be as many as 20 pathways originating within the retina and that each of these pathways communicates its signal to a different location in the central nervous system**. ... We may presume that each subcategory of ganglion cell specializes in communicating about certain types of visual information." Note that there are layers of the following cells as one moves from the photoreceptors to the optic nerves: horizontal, bipolar, Müller cells, amacrine, and ganglion. See Figure 13. "The precise site where information is segregated into different visual streams within the retina is not certain, but it seems likely that the segregation begins immediately at the output of the rods and cones. There **appear to be about 15-20 different bipolar cells that make contact with each cone**. The information encoded by each of the bipolars may serve as the starting point for the [various] visual streams."[127] Oyster points out that there are **between 50 and 100 different retinal cell types** for the retina to function correctly.[128]

The retinal neural system is amazingly complex and efficient. It preprocesses retinal information into visual streams that are sent

---

[126] Palmer, p. 31

[127] Wandell, p. 116

[128] Oyster, p. 79

to the visual cortex. Remember that 40-50% of the brain is dedicated to visual processing and that the retina develops from the same cells as does the brain. Consequently, there is an intimate interconnection between the retina and the brain. The visual cortex is the largest system in the human brain. It is highly specialized to perform extremely specific imaging tasks, from detecting the angle of lines to recognizing faces. Other types of image processing include edge detection, depth perception, color perception, motion perception, object recognition, and dozens of others. We have barely touched on image processing in the brain, nor is it our purpose to do so. Rather, the goal is to introduce the idea that there is an even more complex system, the eye-brain system. The eye-brain system is also irreducibly complex and perfectly functionally coherent.

Hopefully, we have come to realize the complexity of the human eye and its marvelous design. We have addressed the evolutionists' claims that, in certain aspects, the eye is poorly designed. **Every presumed error has, instead, been shown to be an intricate design detail that improves the eye's overall imaging ability**. No less a scientist than Isaac Newton "argued for the intelligent design of the human eye in his classic work, Opticks[129]" which analyzes the fundamental nature of light and is "considered one of the great works of science in history."[130] This section should demonstrate that, contrary to what evolutionists think, optical engineers, vision scientists, neuroscientists, and practicing optometrists and ophthalmologists know that the human eye is a superbly designed optical system. QED![131]

---

[129] Newton, Opticks: Or a Treatise of the Reflections, Refractions, Inflections & Colours of Light, based on the 4th Ed., London, 1730; Dover Books, 2012

[130] Meyer, The Signature of the Cell, p. 145

[131] Latin phrase, 'quod erat demonstrandum,' meaning 'the proof exists.'

If Darwinian evolution cannot explain more than simple adaptations within specific organisms, it is woefully inadequate to explain the Holy Grail of evolution – the origin of life. The origin of life is the subject of the next section. While this topic could fill volumes, we will just take a cursory glance, enough so that we can better understand a few of the major issues.

The Origin of Life

The neo-Darwinian theory applies to living organisms. However, evolution presupposes an ultimate ancestor – the first living organism. According to evolutionists and naturalists, everything in nature must have a natural explanation, i.e., it must be due to raw physical or chemical action. Consequently, the first living organism must have arisen from inert matter. **The ultimate question is how life could have spontaneously arisen from non-living matter?** The origin of life is the ultimate evolutionary challenge. This problem is far more difficult than ordinary evolution. Additionally, traditional evolution does not apply here. There can be no mutations since there is nothing to mutate. Before life and the ongoing struggle to live, there can be no natural selection. Origin of life researchers must ask themselves very fundamental questions. "**One of the most obvious big questions – how did life arise from inorganic matter? – remains a big unknown.**"[132] [emphasis added]

It is presumed that Harold Urey and a graduate student, Stanley Miller, at the University of Chicago in 1953, started the modern era of experimental research into the origin of life. This was 67 years ago as of this writing. They put gas in a glass apparatus containing inorganic compounds that they thought represented the atmosphere of the early earth, applied heat, and sent electrical sparks through the gas. Urey and Miller were amazed to observe

---

[132] American Scientist, Vol. 97, No. 3, May-Jun2009, *The Origin of Life*, Trefil, Morowitz and Smith, p. 206; https://www.americanscientist.org/article/the-origin-of-life

the creation of simple amino acids – organic compounds (containing carbon), the basic building blocks of life. It was the first-time organic compounds were observed to have formed without the presence of enzymes. Later research indicated that their gas did not resemble the earth's early atmosphere, nevertheless, the genie's bottle had been opened.

Since then, scientists, "have found many of the same simple organic molecules in meteorites, comets, and even interstellar gas clouds. ... the simplest of the molecules we find in living systems ... seem to be quite common in nature."[133] A scenario was soon developed, called the Oparin-Haldane conjecture, more popularly known as the 'primordial soup.' "The scenario went like this: After the Earth cooled enough to allow oceans to form, the Miller-Urey process or something like it produced a rain of organic matter. In a relatively short time, the ocean became a broth of these molecules, and **given enough time**, the **right combination of molecules came together** by ***pure chance*** to **form a replicating entity of some kind that evolved into modern life**."[134] [emphasis added] Here we go again – lots of time, pure chance, an entity of some kind, evolved - recall that things could not 'evolve' yet.

The *Origin of Life* article continues, "the essential legacy of the Primordial Soup was twofold: It simplified the notion of the origin of life to a **single pivotal event** [some kind of replicating molecule suddenly came alive] and then it proposed that this event — was a ***result of chance***. In the standard language, ***life is to be seen, in the end, as a 'frozen accident.'*** **In this view, many fundamental details about the structure of life are NOT AMENABLE TO EXPLANATION.**

[133] American Scientist, Vol. 97, No. 3, May-Jun2009, *The Origin of Life*, Trefil, Morowitz and Smith, p. 206

[134] American Scientist, Vol. 97, No. 3, May-Jun2009, *The Origin of Life*, Trefil, Morowitz and Smith, p. 206

The architecture of life is JUST ONE OF THOSE THINGS." [135] [emphasis added] Wow! And Wow! It just happened! We do not know how, or when, or why – it just happened somehow! A 'frozen accident' is a single, one-time, chance event that changed the future permanently. Trefil, Morowitz, and Smith do comment that "although many modern theories are less extreme than this, frozen-accident thinking still influences what some of us ask about the origin of life and how we prioritize our experiments."

Origin of life researchers have been seeking answers for nearly 70 years now. A major problem with early models was the chicken and the egg problem; which came first, proteins or DNA? As we discussed earlier, life requires both DNA and proteins. An early model proposed that proteins came first. This was soon replaced by the DNA-first model. One of the insurmountable problems with the origin of life is the origin of information-rich DNA. Where did the information come from? In the 1980s, an RNA world model gained favor and is still the most popular research model. There is a multitude of problems associated with the origin of life models. We will settle for briefly introducing you to three of them: the RNA World, the origin of biological information, and thermodynamics.

An RNA World

As previously discussed, modern life is primarily based on DNA for genetic information and encoding proteins, life's workhorses. For life to perpetuate itself, it must replicate its DNA. One of the principal tasks that proteins perform is to catalyze [initiate or speed up] chemical reactions. A protein that acts as a catalyst is referred to as an enzyme. According to the origin of life researchers, RNA is the only known chemical that can both act like enzymes and self-replicate. Consequently, the RNA World became their great hope.

---

[135] American Scientist, Vol. 97, No. 3, May-Jun2009, *The Origin of Life*, Trefil, Morowitz and Smith, p. 206

Chapter 14 of Steven Meyer's book, Signature in the Cell[136] is a good reference. See Figure 15. As we walk our way through the proposed steps in the RNA World Scenario, think about the probability of these steps happening by pure chance.

The 7 steps shown in Figure 15 are as follows: (1) simple organic compounds are formed on the early earth – the prebiotic soup; (2) RNA building blocks link up into small polynucleotides – strings of the molecules used to build RNA, i.e., ribose, the four bases, and phosphate groups; (3) an RNA replicase ***arises by chance*** and ***selective pressures occur*** favoring their **increase in complexity**; (4) RNA enzymes (ribozymes) begin **synthesizing proteins** from RNA templates; (5) **eventually, protein-based protein synthesis replaces RNA-based protein synthesis**; (6) reverse transcriptase ***transfers genetic information*** from RNA to DNA; (7) the **modern gene expression system arises** within a **proto-membrane.** [137] Goodness! ***There is a lot of faith involved in the above RNA World scenario***. These steps were described with the following terms: chance, selective pressure, increase in complexity, begin synthesizing proteins, reverse transcriptase, genetic information, gene expression system arises, and proto membrane.

[136] Meyer, Signature in the Cell – DNA and the Evidence for Intelligent Design, 2009

[137] Meyer, Signature in the Cell, Fig. 14.2, p. 299

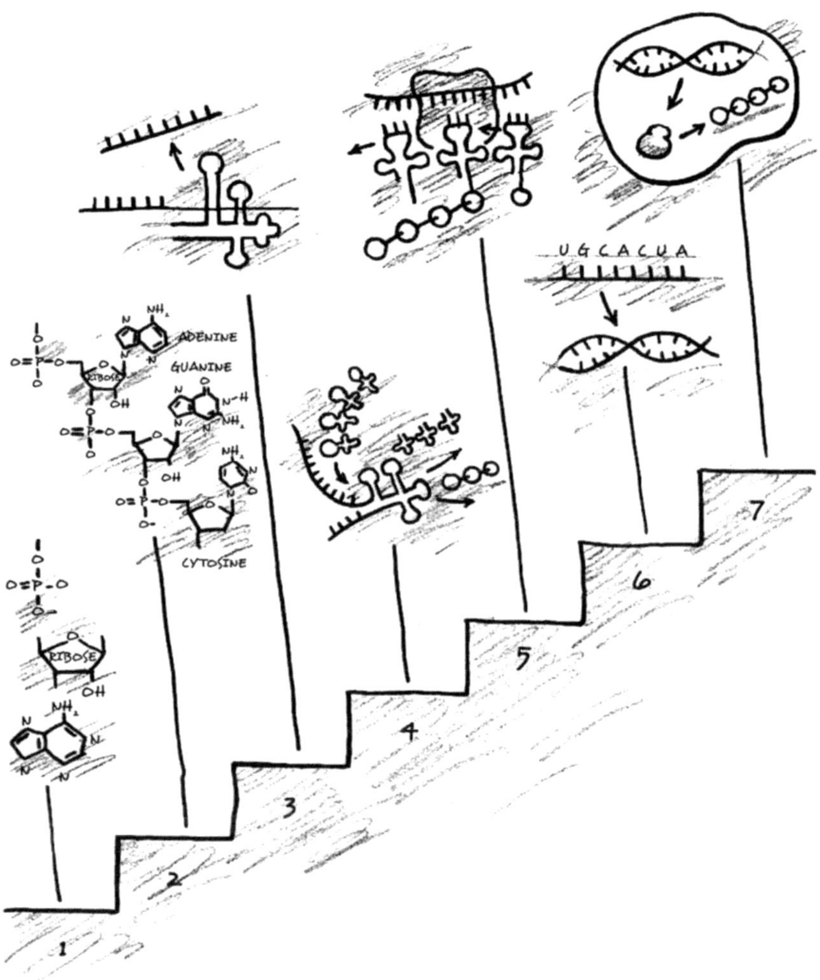

FIGURE 15. THE 7 STEPS IN THE RNA WORLD SCENARIO. [138]

Each of the above conjectures deserves some comments. We have calculated the virtual improbability of randomly encoding a protein. Why would the chance of the random appearance of an RNA replicase be any more probable? Remember how intricate, large, and complicated the replisome was? It is hard to imagine what kind of selective pressure is supposed to occur. Without life and the survival of the fittest, there is no natural selection

[138] Illustration ©2019 Ray Braun; used by permission (Ray Braun Design, Shoreline, WA)

according to Darwin's theory – any pressure must be strictly chemical. As for an increase in complexity, there is no known driver in nature towards increased complexity. Thermodynamics favors the descent into chaos and disorder (see the thermodynamics section below). We have seen that after encoding, synthesizing proteins require incredibly complicated molecular machines, aka spliceosomes and ribosomes. Where do they come from in this scenario? More improbably, where did the information come from to begin with? No random, chance process has ever been discovered or demonstrated that can create useful information. In normal cells, we have seen that a gene expression system involves genes, replisomes, spliceosomes, editosomes, and a host of complicated processes. It seems unlikely that a gene expression system would simply 'arise.' While a simple fatty acid membrane is easy to create, cell membranes are in and of themselves extremely complicated. Cell membranes are composed of phospholipids arranged so that the hydrophilic ('water-loving') heads of the molecule are on the inside of the cell and the hydrophobic ('water-hating') tails on the outside. Scattered throughout the membrane are numerous types of proteins forming channels, pores, and gates. Carbohydrates are also found in the membrane, mostly glycoproteins which attach to certain proteins. Functionally, cell membranes provide a physical barrier, possess selective permeability, must permit the inflow and outflow of large substances, and participate in cell-to-cell signaling. There are a whole bunch of 'ifs' in the RNA World scenario.

The following is a summary found at the end of the chapter entitled *The RNA World and the Origins of Life* in one of the highly used college-level textbooks on cell biology, Molecular Biology of the Cell.[139]

---

[139] Alberts B., Johnson A., Lewis J., et al, Molecular Biology of the Cell, 4th Edition, Garland Science (2002); extract found on the NCBI Bookshelf, a service of the National Library of Medicine, National Institutes of Health.

> "... it **seems likely** that the development of the directly autocatalytic mechanisms [RNA acting as a catalyst facilitating other RNA chemical reactions] fundamental to living systems began with the **evolution of families of molecules** that could catalyze their own replication. **With time**, a family of cooperating RNA catalysts **probably** developed the ability to direct synthesis of polypeptides. DNA is **likely** to have been a late edition: as the accumulation of additional protein catalysts **allowed** more efficient and complex cells to **evolve**, the **DNA double-helix replaced RNA** as a more stable molecule for storing the **increased amounts of genetic information** required by such cells."

Gee-whiz! Another 'just so' story. Words and phrases include: seems, likely, the evolution of molecules, with time, a family of cooperating catalysts, probably, likely, the accumulation, more complex, and increased amounts of genetic information. The leaps of faith in the above summary are plentiful and naïve. This is true of all Origin of Life stories. You can now recognize these 'just so' stories!

Meyer summarizes five problems he has identified with the RNA World scenario. Problem 1: RNA building blocks are hard to synthesize and easy to destroy; RNA does not survive well in water. Problem 2: Ribozymes (RNA molecules that act like protein enzymes or catalysts) are poor substitutes for proteins - "to date, scientists have shown that RNA catalysts or ribozymes can perform only a small handful of the thousands of functions performed by modern proteins". Problem 3: An RNA-based translation and coding system is implausible (because of the irreducible complexity of the DNA coding, transcription, and translation system). Meyer details a 'simple' bacterial cell's translation and coding system. It consists of 106 distinct but functionally integrated proteins along with mRNA, tRNA, and rRNA. These occur inside a ribosome composed of 50 different proteins, 20 distinct tRNA synthetases,

20 distinct tRNA molecules, various additional proteins, free-floating amino acids, ATP molecules for energy, and information-rich mRNA transcripts. Problem 4: The RNA-World does not explain the origin of genetic information. Problem 5: Modern ribozyme engineering does not simulate undirected chemical evolution. The issue of investigator interference when conducting experiments was briefly mentioned before. "Because of the difficulties with the RNA-World hypothesis and the limited number of enzymatic functions that naturally occurring ribozymes can perform, a new cottage industry has sprung up in molecular biology."[140] Their goal is to improve the functionality of existing ribozymes or create new ones that can function more like proteins do. This effort, while scientifically useful, is heavily directed and not in any way representative of how undirected ribozyme evolution would need to operate.

The origin of life 'cottage industry' has been researching and conducting experiments now for over 50 years. Their goal has been to create an RNA World to prove their conjectures. They have published some remarkable work about RNA which may have medical benefits in the future, however, they have not created anything that even resembles a single functional proto-cell.

The Origin of Biological Information

Meyer references Dr. Seuss's children's book, The Cat in the Hat Comes Back. He recalls that one of the cats in the story produces a mysterious device called *Voom*, which cleans up every shred of pink which has accumulated in the house. "Origin-of-life researchers have been looking for their 'Voom' for over fifty years, a process that can, once and for all, clean up the problem of explaining the origin of information." Continuing, Meyer says, "**I discovered that every attempt to explain the origin of biological information either failed or transferred the problem elsewhere** – by presupposing some other unexplained sources of information or

[140] Meyer, Signature in the Cell, p. 301-321

by overlooking the indispensable role of an intelligence in the generation of the information in question."[141] Even within living organisms, "**there are no examples of mutations that add any information to the genome**,"[142] The probabilities of creating even a single small protein let alone a minimal cell organism were discussed previously; the probability is essentially zero. So how are completely random chemical processes supposed to produce the slightest bit of useful information? The theory of thermodynamics implies that it is folly to assume random processes can create ordered information. Thermodynamics will be discussed next.

Thermodynamics

On one hand, the concepts of thermodynamics can be difficult to understand. On the other hand, Albert Einstein exclaimed, "classical thermodynamics ... is the only physical theory of universal content which I am convinced ... will never be overthrown."[143] This is because thermodynamic systems are composed of particles whose average motions define the systems' properties without requiring detailed knowledge of the particles themselves. Think of a cup of hot water on the kitchen counter that gradually cools down. The water was hot because, on average, the water molecules were moving faster than in cold water. The cup gives off heat to the kitchen environment and over time it cools down, i.e., the water molecules collectively slow down. The air in the kitchen is now ever so slightly warmer. Instead of water, the same phenomenon would happen to a glass of warm milk – or any other warm liquid (the type of particle does not affect the outcome). One of the primary applications of thermodynamics in modern science pertains to chemical reactions. Unfortunately for evolutionists,

---

[141] Meyer, Signature in the Cell, pp. 271-272

[142] Spetner, Not By Chance!, p. 173

[143] Martin J. Klein, *Thermodynamics in Einstein's Thought*, Science, Vol 157, Issue 3788, 04 August 1967, p. 509

thermodynamics presents a huge problem for the kinds of chemical reactions that are required for the origin of life.

To understand thermodynamics, we need to understand a few definitions. A thermodynamic system is a system composed of matter and energy, e.g., the primordial soup discussed previously. The entire earth is a thermodynamic system as is your home and your automobile. There are three kinds of systems: (1) isolated; (2) closed; and (3) open. An isolated system does not exchange energy or matter with its surroundings. A closed system can exchange energy (heat or work) but no matter. An open system can exchange heat, work, and matter with its surroundings. The earth is essentially a closed system. While there is some mass exchange with space (cosmic dust falls to the earth, atmospheric gases escape into space), the amount of mass exchange is negligible. **The earth does receive massive amounts of energy from the sun and is therefore a closed system**. The earth is not an isolated system. Also, thermodynamic systems may be in equilibrium or not. Equilibrium means that the system is not changing or only changing very slowly. The small water cup cooling down slowly in a large kitchen is close to being in equilibrium. It would be quite another thing if one fired up a kerosene heater in the middle of the kitchen. The physics and mathematics of thermodynamic systems have only been developed for systems in or near equilibrium.

There are four laws of thermodynamics (0th, 1st, 2nd, and 3rd).[144] For our purposes, we are only interested in the 2nd law. The 2nd Law of Thermodynamics states that any isolated system of matter and energy will tend towards disorder and decay. The degree of disorder is defined by a thermodynamic property called entropy. High entropy means the system is highly disordered. The 2nd Law of Thermodynamics is evidenced in the everyday world we live in – things that are unattended always decay, they become more

---

[144] Wikipedia: Thermodynamics; https://en.wikipedia.org/wiki/Thermodynamics

disorganized and less orderly over time. Think of an ice cube on the kitchen counter. With time, it melts and becomes much less orderly. A corollary to the 2nd Law states that any system cannot spontaneously become more orderly, i.e., the melted water on the counter can never spontaneously become an ice cube again. Outside intervention, however, such as the addition of a refrigerator to the water-kitchen system would enable the liquid water to be turned back into ice again (this is an example of negative entropy which cannot occur spontaneously). Evolutionists are desperately "looking for a way to overcome the insuperable barrier... [that] the Second Law of Thermodynamics poses against an evolutionary origin of life."[145]

Evolutionists claim that closed or open systems, where energy enters the system, can overcome the fact that order must always decrease. Life is highly ordered. Unfortunately, even in a closed or open system, nearly everything still tends to become more disorderly. The earth receives energy from the sun daily, yet things always get 'old' and decay. To overcome this problem, **a coupling mechanism of some kind is necessary to channel the incoming energy to increase order and reduce entropy, at least in some small location**.

Nature does NOT tend to become more orderly without an outside intervention like the refrigerator. Life itself needs to be extraordinarily orderly, otherwise, it would cease to remain alive. Examples of disorder in living organisms include cancer (runaway growth of cells), thyroid malfunction (hormonal imbalance), eye cataracts (lens darkening with age), hemophilia (genetic mutation), and the ultimate disorder- death (everything quits working).

As mentioned, entropy is a measure of disorder. At one extreme, crystallized water or ice has extremely high order and extremely low entropy (the water molecules are packed into a rigid

---

[145] Institute for Creation Research, *Thermodynamics and the Origin of Life – Part II*, April 1, 1978, Duane Gish

crystalline structure and are barely moving). Liquid water has medium order and medium entropy (the water molecules are close together and moving slowly). Water vapor has exceptionally low order and exceptionally high entropy (the water molecules are extremely far apart and moving very rapidly in all directions). Using our hot cup of water on a kitchen counter example, the entropy of the water cup goes down as it cools while the entropy of the air in the kitchen goes up. The combined entropy of the water and the air increases ever so slightly.

Life by its very nature has incredibly low entropy; in fact, the totality of living cell processes is designed to maintain order. Once again think of cells, their smaller organelles, DNA, replisomes, spliceosomes, proteins, etc. Consequently, life could not be created nor maintained in an isolated system in or near equilibrium. It is presumed that a closed system like the earth, with energy coming from the sun, would have the capability of decreasing entropy, at least in localized areas. Optimistically, life would arise somewhere. The question is how? "On the theoretical side, we have to start with the realization that if we apply standard equilibrium thermodynamics to living systems, we arrive at something of a paradox. **Living systems possess low entropy, which makes them very improbable from the equilibrium thermodynamic viewpoint**. ... What is needed, therefore, is an extension of ordinary thermodynamics so that it can apply to systems maintained far from equilibrium by the flow of energy." Hopefully, then, they can mathematically prove that they can lower entropy somehow.

Charles B. Thaxton, Walter L. Bradley, and Roger L. Olsen have analyzed biological thermodynamics in detail.[146] In particular, they

---

[146] Thaxton, Bradley and Olsen, The Mystery of Life's Origin, 1984; Thaxton et al, Mystery of Life's Origin – The Continuing Controversy, 2020; In the foreword to the Continuing Controversy (2020), it states that the original text from the 1984 edition is only lightly updated in the 2020 edition where the text is reproduced in full. "The fact that only light updating was needed is a testament to the meticulous scholarship of the

analyzed non-isolated systems. A system that receives continuous energy flow, such as the earth does from the sun, is the equivalent of having continuous work done on the system.[147] Unfortunately for evolutionists, having outside energy does not guarantee that useful work will occur.

Just having the ability to do work on the system is not good enough. **There must be "a coupling mechanism … to allow the released energy to be converted into a particular kind of work**." This coupling mechanism is an analog to our refrigerator turning liquid water into ice, i.e., lowering entropy. Because of the nature of biochemical bonding "the formation and maintenance of living systems … requires continuous work to be done on the system."[148] In plants, energy flow is supplied principally by solar radiation. In animals, the energy flow is provided by eating high-energy biomass. There are two kinds of work needed for biological systems: (1) **chemical work**; and (2) **configurational work**. [149] Previously, the origin of life scientists have focused only on the first kind of work, chemical work, i.e., creating biochemicals. Unfortunately, that is not enough. As we have seen, proteins and DNA/RNA are very informationally rich chemicals in which the highly specific arrangement of molecules (their sequences and their shapes) is critical to their usefulness. Therefore, work must be performed to **arrange molecules in the correct order** and there must be a **mechanism for coupling** that work (think replisome, spliceosome, etc. which we have discussed). Even if the right kind of work can be used to create all the needed biomolecules, ***there is***

---

authors and to the enduring nature of the problem they identified in the origin of life studies." **The problem was that "a profoundly informative intervention" or an "intelligent cause" is needed**. (p. 8)

[147] Thaxton, Bradley and Olsen, The Mystery of Life's Origin, 1984, p. 122

[148] Thaxton, Bradley and Olsen, The Mystery of Life's Origin, 1984, p. 123

[149] Thaxton, Bradley and Olsen, The Mystery of Life's Origin, 1984, pp. 135-141

no known coupling mechanism for doing configurational work outside of already existing life-forms.

From Figure 15, Step 4 of the RNA World Scenario suggests that ribozymes (RNA enzymes) begin synthesizing proteins from RNA templates. The RNA templates must contain accurate and detailed information to form useful proteins. In Step 5, RNA-based protein synthesis must be replaced by protein-based synthesis. Both synthesis systems require numerous biological machines to work. There must be information to properly encode these machines. Step 6 requires a transfer of genetic information from RNA to DNA. Finally, Step 7 requires a complete gene expression system requiring information to encode the entire system. Life requires massive amounts of information. Where did the information come from originally? And if it existed, how did it get coupled into DNA, RNA, proteins, and the like BEFORE configurational work mechanisms existed?

Hope springs eternal for naturalistic origin of life researchers. Armand Delsemme, a professor of astrophysics at the University of Toledo, Ohio, tries to rescue origin of life theories. He speculates that "the **remaining difficulties in explaining the origin of life from inert matter** arise only from **our ignorance**, and not from apparently blind alleys which would all turn out to be dead-ends."[150] How can he say that when he just acknowledged current scientific ignorance concerning the origin of life? Only an Intelligent Designer knows all the laws of chemistry and physics and can supersede them if necessary. **There is not a smooth, continuous progression from inanimate biochemicals to a living organism – it is a step function, an infinite jump from non-life to life**. According to the doctrine of The Church of Jesus Christ of Latter-day Saints, life begins when the spirit body enters the physical body and is enabled by the 'light of Christ.' (often referred to as the 'spark of life;' D&C 88:13; see Chapter 10.) It takes much more faith to

---

[150] Delsemme, Our Cosmic Origins, p. 149

believe that the origin of life came about by a 'frozen accident' than to give the Creator His due.

Human Evolution

Paleoanthropology (the study of ancient humans) is the epitome of what is wrong with the study of evolution. One, it is promulgated as a given and is a direct assault on the existence of God by definition. Two, it is less professional and objective than other evolutionary sciences, let alone physics and chemistry. Third, it is less exacting than most sciences. Fourth, fossil discovers are extremely jealous of their finds and limit access to the fossils by outside parties. Fifth, it is subject to exaggeration and deception. Let us briefly discuss each issue.

A fundamental tenet of evolution is that all life forms descended from the original life form. The fundamental tenet of paleoanthropologists is that man descended from primates. Philip Johnson said, "the story of human descent from apes is not merely a scientific hypothesis, it is the secular equivalent to the story of Adam and Eve and a matter of immense cultural importance."[151]

According to Marvin Lubenow, "paleoanthropology does not deserve the same status in science that is accorded fields like chemistry and physics." Continuing, he comments that "**this field is the scene of much prejudice, subjectivity, and emotionalism in the interpretation of the human fossils and the construction of phylogenic trees**. The professionalism and objectivity found in other areas of science have been conspicuously absent in this area." [152] In his book, Bones of Contention, Lubenow outlines

---

[151] Johnson, Darwin on Trial, p. 83

[152] Lubenow, Marvin L., Bones of Contention: A creationist assessment of human fossils, Baker Books (2004), p. 35; Lubenow is a professor of Bible, theology, and apologetics at Southern California Seminary. He has studied human fossils for over thirty-five years. The first addition of the aforementioned book was in 1992.

several faulty lines of evidence used by evolutionists to promote human evolution.[153] One faulty line of study is the study of living chimpanzees as a way to study what early humans may have been like. This begs the question, however. Without evidence, they have **assumed** that humans descended from chimpanzees. Another fallacy is their failure to understand the difference between scientific and historical evidence. Evolution, especially human evolution, is more of a historical exercise than a scientific one. **Science can show that something is possible, but it cannot prove that something happened in the past** – proof of the past requires some sort of historical evidence. The third and most serious fallacy is the supposed hominid family tree. See Figure 18 in Appendix 2. **This "sequence, impressive as it seems, is a very artificial and arbitrary arrangement."** What the public does not know is that **"(1) some fossils are selectively excluded if they do not fit well into the [foreordained] evolutionary scheme; (2) some fossils are arbitrarily downgraded to make them appear to be evolutionary ancestors when they are in fact true humans; and (3) some nonhuman fossils are upgraded to make them appear to be human ancestors."**[154]

Andrew Hill, a Yale University paleoanthropologist, made some damning remarks about his field of study. "Compared to other sciences, the mythic element is greatest in paleoanthropology." [155] Shortly after making the above observation, he made another rather humorous observation. He said that any casual student of paleoanthropology can easily discern the rules of the field. They are: "(1) every new fossil hominid specimen is the most important ever found and solves all

---

[153] Lubenow, Bones of Contention, pp. 33-34

[154] Lubenow, Bones of Contention, pp. 34

[155] Hill, Andrew, *The Myths of Human Evolution*, a review of The Myths of Evolution by Niles Eldredge and Ian Tattersall, American Scientist, 72 (March-April 1984), p. 189

phylogenetic problems; (2) every new hominid specimen is completely different from all previous ones, no matter how similar; and (3) every new hominid specimen is a new species, and probably a genus, and therefore deserves a name."[156]

To illustrate what limited access means, let us consider the Dead Sea Scrolls. Most people have heard of these. The scrolls are ancient Jewish religious manuscripts that were discovered in the hills overlooking the Dead Sea. Most of the scrolls were discovered after WWII. Whichever lucky scholar got their hands on a particular scroll guarded it zealously. "Publication of the scrolls has taken many decades, and delays have been a source of academic controversy. The scrolls were controlled by a small group of scholars ... while a majority of scholars had access neither to the scrolls nor even photographs of the text." [157] According to Lubenow, access to newly discovered hominid fossils is even more guarded than the Dead Sea Scrolls! One can only imagine why? It would seem to be an effort to prevent independent verification of the conclusions drawn by the discoverers.

Finally, there is the issue of deception through the use of artists' pictures, drawings, and reconstructions. "**No tool has been as successful in promoting human evolution as have been the pictures and reconstructions of our ancient ancestors**. Since no one has ever seen these ancient ancestors, the abilities of the artists who constructed them have been nothing short of miraculous. It gives the term 'science fiction' a whole new meaning."[158] David Reybrouck has studied artistic depictions of presumed human fossils. He said that their **drawings and reconstructions "(1) always go beyond the archeological data; (2)**

---

[156] Hill, *The Gift of Taungs*, review of Hominid Evolution, edited by D. V. Tobias, Nature, 323, (18 September 1986), P. 209

[157] Wikipedia: Dead Sea Scrolls; https://en.wikipedia.org/wiki/Dead_Sea_Scrolls

[158] Lubenow, Bones of Contention, p. 38

always involve speculations and prejudices of the fossil discoverers, who advise the artists; (3) always involve interpretations that are theory laden; and (4) always are nonobjective but are trusted to be accurate; and (5) are used so extensively because they sell evolution so effectively." [159] One of the most well-known is the depiction of a series of upright, walking proto-apes, then apes, then modern man. Paleoanthropologists know that proto-apes and apes have never walked upright. The entire depiction scheme of fossil humans sounds like the deceitful images discussed in the Icons of Evolution section.

A short discussion of the famous Lucy fossil should be enlightening. Lucy was found in Ethiopia in 1974 and is dated to be ~3.2M years old. The left half of Figure 16 depicts Lucy's bones while the right half depicts a reconstruction of those bones at the Cleveland Museum of Natural History.[160] Several major anatomical features that distinguish humans from apes are brain size, the hip, and the ankle, i.e., whether or not the hominid is bipedal and walks upright or not. The first thing to notice about Lucy's skeleton is that it is remarkably complete (~40%) by paleoanthropology standards but is nevertheless incomplete, especially the skull. According to evolutionists, Lucy has an apelike brain and humanlike hips and ankles (bipedal). Part of the hip bone was distorted leading to two opposing reconstructions, one apelike and one humanlike. Notice the complete absence of the front of the skull. Skulls of other presumed Australopithecus hominids must have been used to reconstruct the face and head of the Lucy fossil.

This is more imagination and science fiction than it is science. The principal takeaway from this section is that the human

---

[159] Reybrouck, David Van, *Imaging and imagining the Neanderthal: the role of technical drawings in archaeology*, Antiquity, Vol. 72, No. 275, 1998, p. 63

[160] Wikipedia: Lucy (Australopithecus); https://en.wikipedia.org/wiki/Lucy_%28Australopithecus%29

evolutionary tree is 'artificial and arbitrary.' Note that "any series of objects created by humans (or [even by] God) can be arranged in such a way as to make them look as if they had evolved when in fact they were created independently by an intelligent being. The fact that objects can be arranged in an 'evolutionary' sequence does not prove that they have an evolutionary relationship or that any of them evolved from any of the others."[161]

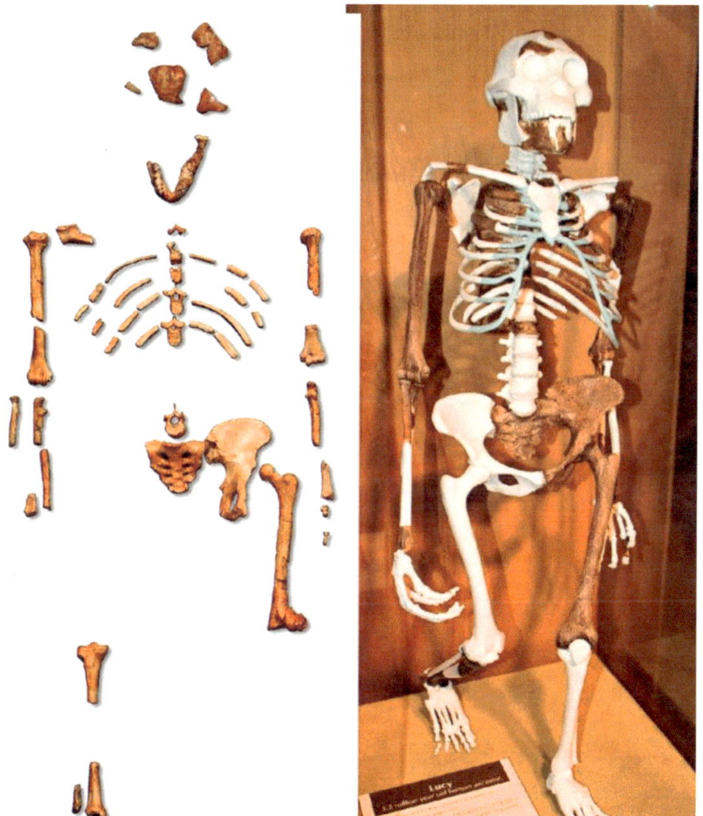

FIGURE 16. BONES AND RECONSTRUCTION OF THE PRESUMED HOMINID FOSSIL LUCY (AUSTRALOPITHECUS). [162]

---

[161] Lubenow, Bones of Contention, p. 35

[162] Wikipedia: Lucy (Australopithecus); left, CC BY 2.5; right, CC BY-SA 2.0; https://en.wikipedia.org/wiki/Lucy_%28Australopithecus%29

8. INTELLIGENT DESIGN AND THE LIGHT OF CHRIST

Intelligent Design (ID) is the proposal that a God-like, intelligent entity was intimately involved in the design of life on earth. ID is somewhat different than creationism in that the days of creation in the Bible are removed from the discussion. However, both worldviews insist on an Intelligent Designer with or without the Bible. We will first hear quotes from some of the scientists and philosophers that are pro-ID followed by the less-than-flattering treatment of ID by those who profess organic evolution. Those criticized include the Discovery Institute, an organization dedicated to promoting Intelligent Design. The chapter will then begin to focus on whether God needed evolution to create organisms, including man.

This chapter will start with a discussion about how finely tuned the universe is for life to exist. It will posit the question, does God need evolution to create life? The answer is a clear 'no.' This will be demonstrated by discussing miracles. Miracles are accomplished by the power of the Priesthood commanding the forces of nature through the Light of Christ. Two modern-day miracles, the First Vision and Moroni's visit to the boy prophet, Joseph Smith, will be discussed because they demonstrate heaven's control over the laws of nature. This will be followed by a review of the scriptures and what they have to say about plants, animals, and humans. The scriptures are truly clear that life-forms reproduce 'after their kind.' Lastly, the official statement of The Church of Jesus Christ of Latter-day Saints referred to as the *Origin of Man* will be summarized followed by some recent quotes by current Apostles. The purpose of this chapter is to explain Intelligent Design, to strongly propose that the **Gods do NOT need evolution** to create life-forms and that the Gods work through the Light of Christ controlling the forces of nature.

For me, the Intelligent Designers are God the Father and His Son, Jesus Christ. See Chapter 10. Another name for the Intelligent Designer or God is The Great Watchmaker. Newton believed "that a single Mind had designed the universe. And surely, if the universe were the work of a single Mind, it would follow a single consistent set of principles. Operating on that assumption, Newton demonstrated that the entire universe is a single uniform system, describable everywhere by the same mathematical laws."[163]

A web search of Intelligent Design usually turns up derogatory comments. For example, the Wikipedia page for 'Intelligent Design' contains the following statements. "**Intelligent design** is a **pseudoscientific argument for the existence of God**, presented by its proponents as an 'evidence-based scientific theory about life's origins.'" "**ID is a form of creationism** that lacks empirical support and offers no testable or tenable hypotheses, so **it is not science**." This assumes that microevolution, macroevolution, and the origin of life have been tested. I think we have shown that the latter two processes have not been tested or verified. "The leading proponents of ID are associated with the **Discovery Institute, a Christian, politically conservative think tank** based in the United States."[164] [emphasis added] Critics of ID deride the ideas of God, creationism, and the Discovery Institute. The Institute is referred to as 'Christian' and 'politically conservative' – as if those things are anti-science and inherently evil. This anti-ID worldview is the ultimate expression of the science of naturalism and represents a naturalistic, atheistic worldview. Below are what some scientists and evolution evangelists have to say about religion, in general, and Christianity, in particular. Before we get to the anti-ID statements, it should be helpful to hear first from some advocates of ID.

---

[163] Nancy R. Pearcey and Charles B. Thaxton, The Soul of Science: Christian Faith and Natural Philosophy, Vol. 16, Crossway Books, 1994

[164] https://en.wikipedia.org/wiki/Intelligent_design

Six pro-ID quotes should be enough, three from Isaac Newton and one each from Rene Descartes, David Gelernter, and Michael Behe. Both Gelernter and Behe were firm believers in evolution until after earning their Ph.D.'s in Computer Science and Biochemistry, respectively. Further research into evolution changed their minds. Isaac Newton was an erudite theologian who always believed in a monotheistic God and creator as did Descartes. Newton wrote more than 1M words on theology – more than he did on science – though almost none of those words were published until after his death. Gelernter is a noted computer scientist who, after reading several ID centric books[165], gave up his belief in Darwinian evolution though not whole-heartedly accepting ID. He noted early in his review of the ID books, that "the exceptional intricacy of living things, and their elaborate mechanisms for fitting precisely into their natural surroundings, seemed to cry out for an intelligent designer long before molecular biology and biochemistry. Darwin's theory, after all, is an attempt to explain 'design without a designer.'"[166] Sir Isaac Newton stated, "don't doubt the Creator, because it is inconceivable that accidents alone could be the controller of this universe."[167] "The regular motion of the planets made it reasonable to believe in the continued existence of God." Sounds nearly identical to Alma 30:44, "all things denote there is a God; yea, even the earth, and all things that are upon the face of it, yea, and its motion, yea, and also all the planets which move in their regular form do witness that there is a Supreme Creator." Newton also upheld the idea that "like a ***watchmaker***, God was forced to intervene in the universe and

---

[165] Stephen Meyer, Darwin's Doubt, 2013; David Berlinski, The Deniable Darwin and Other Essays, 2009; David Klinghoffer, Ed., Debating Darwin's Doubt, 2015

[166] Gelernter, *Giving Up Darwin*, Claremont Review of Books, 2019

[167] https://quotesgram.com/img/isaac-newton-quotes-about-god/9384486/

tinker with the mechanism from time to time to ensure that it continued operating in good working order." Like Newton, René Descartes viewed "the cosmos as a **great time machine** operating according to fixed laws, **a watch created and wound up** by ***the great watchmaker***."[168] [emphases added] Michael Behe, whom we have referenced numerous times, said, ***"when one starts to treat Darwinism as a hypothesis about the biochemical level of life rather than an assumption, it takes about ten minutes to conclude it's radically inadequate. It takes perhaps another ten minutes to realize that the molecular foundation of life was designed."***[169]

The Discovery Institute

One of the primary organizations advocating intelligent design is the Discovery Institute's Center for Science and Culture.[170] As already mentioned, mainstream evolutionists revile the work of the Institute. In referring to advocates of ID, Zimmer says they use "the **old attacks on missing links ..., biochemistry and genetics**. They claim that there is **irreducible complexity** to the molecular working of life. To clot a wound, for instance, a long cascade of chemical reactions creates the clotting molecules. Take out one of the elements in the cascade, and a person bleeds to death. How, then, could evolution have built up this system from simpler parts?"[171] At least Zimmer's attack of ID advocates is accurate, they do refer to the problems with missing links (transitional forms), the extreme complexity of biochemistry and genetics, and irreducible complexity – all of which we have discussed previously.

The following attack is more direct. Joseph Kuhn, a surgeon at the Baylor University Center, gave a paper entitled *Dissecting*

---

[168] https://en.wikipedia.org/wiki/Watchmaker_analogy

[169] Behe, Darwin Devolving, p. 8

[170] https://www.discovery.org/id/

[171] Zimmer, Evolution, p. 325

Darwinism during the school's medical proceedings.[172] During his paper, he summarized the ID arguments used before the Texas State Board of Education regarding teaching evolution and alternatives, such as ID. He was mercilessly attacked from multiple fronts. The pro-ID arguments discussed were the inability of evolution to explain the origin of life, irreducible complexity, and the lack of transitional fossils. Jerry Coyne, former President of the Society for the Study of Evolution and editor of Evolution and The American Naturalist, published a review of Kuhn's article. He commented that Kuhn's article was "merely a cobbled-together list of canards from the Discovery Institute" and "he's simply a mouthpiece for the Discovery Institute."[173]

Evolutionists generally ignore the Discovery Institute out-of-hand. However, those associated with the Institute are biochemists, physicists, mathematicians, biologists, medical doctors, lawyers, historians, and the like with credentials every bit as impressive as evolutionists. The Institute is ignored because they do not conform to the prevailing worldview of naturalism within modern science. They are not ignored because of their arguments; their arguments are dismissed out of hand. This is more common in science than you have been led to believe. When I was a teenager, I remember that the concept of plate tectonics came to be accepted as the prevailing theory of how the earth's continents move. Before plate tectonics, the earth's crust was deemed to be static and immovable. A booklet entitled *Continents Adrift: Readings* was published by Scientific American (first published in 1952) which outlined the theory of plate tectonics. The Preface contains the following statements. These articles "describe the latest scientific revolution," that "was long in the making." At a conference of geophysicists in 1971, "it was made clear that the

---

[172] Baylor University Medical Center Proceedings. 25 (1), 2012, 41-47

[173] Jonathan Wells, *"Shut Up," Coyne Explained*, Evolution News, January 19, 2012; https://evolutionnews.org/2012/01/shut_up_coyne_e/

notion of continental drift, which had been ***heresy*** only a few years before, had become the ***orthodoxy*** of the great **majority**." [174] [emphasis added] This more resembles a cultural or religious revolution than a scientific revolution. New thinking had been rejected, and the truth was established by a majority. After all, scientists are people who can also be influenced by human emotions more than most are willing to admit. In modern-day science, scientists can also be driven by funding. We can only hope that today's heresy, Intelligent Design, will in a few years become the future orthodoxy. The advancement of the biological sciences should significantly improve.

Fine-tuning of the universe

Life as we know it is exceedingly sensitive to certain fundamental physical constants (fixed mathematical values) found in the universe. Two examples would be the magnitude of the electric charge of the electron and the ratio of the masses of the proton and the electron. Protons, neutrons, and electrons make up atoms. Stephen Hawking noted, "The laws of science, as we know them at present, contain many fundamental numbers, like the size of the electric charge of the electron and the ratio of the masses of the proton and the electron. ... The remarkable fact is that the values of these numbers seem to have been very finely adjusted to make possible the development of life."[175]

Intelligent Design advocates would insist that the Intelligent Designer knowingly set the values of the constants. Currently, there are about 25 'constants' arising from the standard model of sub-atomic particle physics (particles smaller than atoms like neutrinos, muons, and quarks) and one from general relativity (which describes gravity). Researchers who model the

---

[174] Continents Adrift, W. H. Freeman, 1972

[175] Hawking, Stephen, A Brief History of Time, Bantam Books, 1988, pp. 7, 125

development of the universe find that the nature of the universe is extremely sensitive to the exact values of many fundamental constants. It is not clear which of the 25 or so constants are truly fundamental. On the atomic side of things, the constants affect the way atoms and molecules form or do not form. On the cosmological side of things, they affect the way stars and galaxies form. Both types of constants are critical.

Cosmologically, a good example is the cosmological constant, Λ. This constant is associated with the general theory of relativity (gravity) and has a value of ~10^{-122}. It is so small that it only affects cosmic structures larger than ~1 B light-years across (our Milky Way galaxy is only ~100K light-years across or 10,000X smaller). Before 1998, the assumed value was equal to zero. It was then discovered that the expansion rate of the universe is ever so slightly accelerating. If Λ is too large, the universe would expand too fast for galaxies to form, i.e., there would be no life to begin with. If Λ is negative, the expansion rate of the universe is slowing down and would eventually collapse, i.e., all life would end in a Big Crunch.

Atomically, a good example is the balancing of forces to create stable atoms. The electromagnetic force, one of the four known forces of nature, attracts electrons (- charge) to nuclei (+ charge) to make atoms. This force also causes the protons inside the nucleus to repel each other. The strong force, one of the other forces of nature, causes the protons to be attracted to each other. The strong force is stronger than the electromagnetic force and, consequently, holds atomic nuclei together. Mathematical analysis shows that the magnitude of the strong force must be within 0.5% of its current value and the electromagnetic force must be within 4% of its value to form stable nuclei and atoms.

Cosmologists differ as to why they think the universe is so finely tuned for life. Many say that there must be many universes (the theory of multiverses). We are alive because, in this universe, the constants were 'just right.' A few acknowledge the Creator's hand in purposely setting up the universe in advance.

Did God Need to Use Evolution to Create Life?

Of all religious sects, members of the Church of Jesus Christ of Latter-day Saints should take the scriptures the most literally. Some Christian theologians have rejected many stories in the Bible as mere allegories. There are numerous examples of amazing miracles in the scriptures, including those of Enoch, Moses, Joshua, Jesus Christ, Joseph Smith, the Flood, and the Division of the Earth. We will concentrate on three miracles wrought by Enoch, Moses, and Joshua.

Speaking of Enoch, the Lord said, "And so great was the faith of Enoch that he led the people of God, and their enemies came to battle against them; and he spake the word of the Lord, and **the earth trembled**, and **the mountains fled**, even according to his command; and the **rivers of water were turned out of their course**; and **the roar of the lions was heard out of the wilderness**; and all nations feared greatly, so powerful was the word of Enoch, and so great was the power of the language which God had given him. There also **came up a land out of the depth of the sea**, and so great was the fear of the enemies of the people of God, that they fled and stood afar off and went upon the land which came up out of the depth of the sea."[176] At the beginning of the Israelite Exodus out of Egypt, "Moses stretched out his hand over the sea; and the Lord **caused the sea to go back** by a strong east wind all that night, and **made the sea dry land**, and **the waters were divided**. And the children of Israel went into the midst of the sea upon the dry ground: and the waters were a wall unto them on their right hand, and on their left."[177] Later, in the promised land, Joshua led Israel against the Amorites; with the Lord's assistance, Israel prevails. "The Lord **cast down great stones from heaven** upon them ... then [Joshua] said Sun, stand thou still [and the] Moon, ... And **the sun stood still, and the moon stayed**, until the people had avenged

---

[176] Moses 7:13-14

[177] Exodus 14:21-22

themselves upon their enemies. ... And there was no day like that before it or after it, that the Lord hearkened unto the voice of a man: for the Lord fought for Israel. "[178]

The miracles of Enoch are found in the Book of Moses, a modern revelation. The parting of the Red Sea is mentioned three times in the Book of Mormon, once in the Book of Moses, and once in the Doctrine & Covenants. The Book of Moses is an extract from Joseph Smith's inspired translation of the Bible referred to as the Inspired Version or the Joseph Smith Translation (JST). Joshua's miracle is alluded to in Helaman 12:14-15. Are these events and miracles cited in the scriptures real events or allegories? Are they historical facts or storytelling? Since these events have been confirmed by modern revelation (JST Bible, Book of Mormon, Doctrine and Covenants, and the Book of Moses), members of the Church of Jesus Christ of Latter-day Saints should be willing to accept that the scriptures say what they say and mean what they mean. Miracles occur. We call them miracles because we humans cannot do them unaided by heaven nor can we explain the details of how they happen. Does God need a blind, mindless, purposeless, trial and error process to work the miracle of life? I think not.

This leads us to another question. How does God work miracles? Several sections of the Doctrine & Covenants are insightful, specifically sections 88 and 131.

D&C 88:6-13

> 6 [Jesus Christ is] He that ascended up on high, as also he descended below all things, in that he comprehended all things, that he might be **in all and through all things**, the light of truth;
>
> 7 Which truth shineth. This is **the light of Christ**. As also he is in the **sun**, and the light of the sun, and the **power thereof by which it was made**.

---

[178] Joshua 10:11, 13-14

> 8 As also he is in the **moon**, and is the light of the moon, and the **power thereof by which it was made**;
>
> 9 As also the light of the **stars**, and the **power thereof by which they were made**;
>
> 10 And the **earth** also, and the **power thereof**, even the earth upon which you stand.
>
> 11 And the **light which shineth**, which giveth you light, is through him who enlighteneth your eyes, which is **the same light that quickeneth your understandings**;
>
> 12 Which light proceedeth forth from the presence of God to **fill the immensity of space**—
>
> 13 The light which is **in all things**, which ***GIVETH LIFE TO ALL THINGS***, which is the **law by which all things are governed**, even the power of God who sitteth upon his throne, who is in the bosom of eternity, who is in the midst of all things.

D&C 131:7

> 7 There is **no such thing as immaterial matter. All spirit is matter**, but it is more fine or pure, and can **only be discerned by purer eyes**;

Let us summarize the doctrines contained in D&C 131. We learn that spirit matter, which we mortals cannot sense, is real matter. Science has no idea that spirit matter exists unless it is somehow linked to the recent discovery of dark matter or vacuum energy. Astronomers estimate that as much as 85% of the matter in the universe is composed of dark matter. It is called dark because it is undetectable by current scientific instruments. Its presence can only be inferred from indirect measurements such as gravitational effects.

From D&C 88 we learn that the **Light of Christ** is the key to the creation of the universe and life. Christ comprehended all things, i.e., the Light of Christ permits him to be in all and through all things. Another name for the Light of Christ is the Light of Truth.

Keep in mind that the definition of "truth is knowledge of things as they are, and as they were, and as they are to come."[179] The Light of Christ is in the sun, moon, stars and the earth, and the power by which they were made. **This light is also the law by which all things are governed; this light emanates from the presence of God to fill the immensity of space**. Interestingly, the same light which governs the physical universe is also the light that enlightens men's eyes and quickens their understandings. Most importantly, for our purposes, this ***Light of Christ gives life to all things*** **(D&C 88:13)** – not physical and chemical forces combined with chance and frozen accidents!

Levels of Light

A model of the levels of light should prove useful. The starting inspiration, though less comprehensive, came from Hyrum Andrus' book, <u>God, Man and the Universe.</u> [180]

LEVELS OF LIGHT	
INFLUENCE ON	**INFLUENCE**
Church of the Firstborn	Jesus Christ, the Second Comforter
Members of the Church of Jesus Christ	Holy Ghost, the First Comforter
All Humans	Light of Christ, our Conscience
All Animals	Instinct
Most Plants	Physical Light
Inanimate Matter	Forces of Nature
Spirit Matter	Some/All Traditional Forces plus Some New Forces of Nature?

TABLE 1. THE PROPOSED LEVELS OF THE LIGHT OF CHRIST

Table 1 details the levels of inspiration or governing powers available here in the Telestial World. We will begin in the middle of the table with humans and work ourselves up. The scriptures tell us that all men and women receive the Light of Christ at birth. It is a gift from God to aid our sojourn here in mortality. We call this

[179] D&C 93:24

[180] Andrus, <u>God, Man, and the Universe</u>, 1968, p. 257

influence our conscience. If one desires 'to be a greater follower of righteousness' (Ab1:2) and is baptized and confirmed into The Church of Jesus Christ, they receive a greater light, the Holy Ghost – the third member of the Godhead. The Holy Ghost is also referred to as the Comforter, technically, the First Comforter. If one so lives their life, demonstrating that they are willing to serve God and build the Kingdom of God on Earth despite all hazards, they may receive the Second Comforter. The Second Comforter is Jesus Christ, himself.

Working downward in the table, we come to animals. Animals have instinct. This instinct is given to them to order their lives, to improve their chances of survival, and to enable them to have joy. If humans so sear their conscience by ignoring its promptings, they will descend to the influence of instinct – survival of the fittest, every man for themselves, etc. This is probably what the scriptures mean when they talk about men 'hardening their hearts,' e.g., 1 Ne 17: 30, 41-42. The next level down is physical light which provides life to plants. Below plants is inanimate matter which is influenced by the forces of nature. Beyond mortality, both spirit matter and spirit beings (humans, animals, plants, and matter) must be influenced by similar levels of the Light of Christ.

In modern physics, light particles, or photons, are essential in how electromagnetism works. Electromagnetism is one of the four forces of nature. Photons are the 'force carrier particles' for the electromagnetic force. There are three other forces of nature, each having their own force carrier particles. I believe that the 'light' that governs matter not only means photons but all the force carrier particles.

There are four forces of nature: (1) electromagnetism; (2) strong force; (3) weak force; and (4) gravity. We have briefly touched on (1), (2), and (4). Gravity is the weakest force. If we assign the strength of the force of gravity as 1, the weak force has a strength of $\sim 10^{32}$, electromagnetism a strength of $\sim 10^{36}$, and the strong force is $\sim 10^{38}$. As a review, the atom provides a good demonstration of how the strong and electromagnetic forces operate to create atoms. The atomic nucleus is composed of protons (positively charged) and neutrons (no charge, neutral). The protons repel each other through the electromagnetic force (like charges repel). Nevertheless, the strong force holds the nucleus together since it is much stronger than the electromagnetic force (137 to 1). Negatively charged electrons are attracted to the positively charged nucleus (opposite charges attract) and form a cloud around the nucleus, forming atoms. The weak force is involved in radioactive decay such as carbon-14 turning into nitrogen-14.

In the modern Standard Model of Particle Physics, each force has one or more force exchange particles. Force exchange particles are also referred to as force carrier particles. The force carrier particle for the electromagnetic force is the photon, a light particle. A crude analogy of how the repulsive electric force works through exchange particles can be envisioned as two ice skaters gliding towards each other. Just before physically colliding, each skater throws a heavy medicine ball towards the oncoming skater. Upon throwing the ball, the skater is forced backward from the throwing motion (Newton's equal and opposite action). Furthermore, when each skater catches the incoming ball, they are also driven backward (transfer of momentum). Envision the throwing and catching happening instantaneously. The balls, then, resemble force carrier particles though the particles themselves have no mass. It appears from the outside that the two skaters collided and bounced off each other but, in reality, they never came into contact. Protons and neutrons found in the atomic nucleus are comprised of subatomic quarks. Force carrier particles, gluons, are continuously exchanged between the quarks to produce the strong force and keep the protons and neutrons together. For completeness, the force exchange particles for the weak force are W and Z bosons. The graviton is the theoretical exchange particle for gravity though it has never been physically detected.

It seems logical that the light spoken of in D&C 88:13 "which is the law by which all things are governed" can be extended from only light particles to all of the four force exchange particles. Photons, or physical light particles, are the force exchange particles for the electromagnetic force. To govern all things would require the control of all the forces of nature. To control the forces of nature requires the control of the force exchange particles. Consequently, it makes sense that in this context, 'light' represents all the exchange particles.

Pertaining to spirit matter, the same four forces may be operative or perhaps other forces are manifested. It is reasonable to suppose that these forces are also extensions of the Light of Christ. **Since the Priesthood may exercise control over the Light of Christ as permitted by the Lord, the Priesthood may control all the forces of nature. This fact explains how all miracles and miraculous events occur**.

All the forces of nature operate in consistent ways. We call these natural laws. Science often takes the existence of natural laws as a given. However, Denton[181] claims, "The origin of natural laws that generate the primal order is, of course, its own important question."[182] Whether or not the natural laws are a given or, they too, are designed, the Gods utilize the Light of Christ to design, maintain, and control these laws.

Two excellent examples of how the priesthood can control the forces of nature are found in the appearance of God the Father and

---

[181] Michael Denton graduated from Bristol University as a medical doctor and earned a Ph.D. from King's College in biochemistry. Though a self-described agnostic, he is a strong defender of an anti-Darwinian evolutionary position. (https://en.wikipedia.org/wiki/Michael_Denton) Two of his books are Evolution: A Theory in Crisis – New Developments in Science Are Challenging Orthodox Darwinism, 1985, and Evolution: Still A Theory in Crisis, 2016.

[182] Denton, Evolution: Still A Theory in Crisis, 2016, p. 18

His Son, Jesus Christ, to the boy prophet, Joseph Smith (the First Vision), and the appearances of the Angel Moroni three years later. As you read the Prophet's descriptions of these first visitations, concentrate on the Godhead's and the angel Moroni's control of the forces of nature.

The Appearance of God the Father and His Son, Jesus Christ, to Joseph Smith in the Spring of 1820

In the Spring of 1820, at the ripe old age of fourteen, Joseph Smith went into a grove of trees near his home to offer a prayer. In response to his earnest prayer, he received a visitation from both God the Father and His Son, Jesus Christ. This is referred to as the 'First Vision.' Joseph went into the grove as a boy; he emerged as The Prophet. There are ten accounts of the First Vision. The following is a harmonized account created by Matthew B. Christensen.[183] Christensen uses *Joseph Smith – History* (JS-H) as the running account with inserts from the other nine accounts, as appropriate. The following is the harmonized version of JS-H 1:16-17. The JS-H account is underlined. Emphasis highlights the visitors' complete control of light itself and the forces of nature

"I saw a **pillar of light** [or] fire exactly over my head, which exceeded the brightness of the dazzling sun in his meridian splendor, [and it] **came down from above**, which at first seemed to be a considerable distance. As it drew nearer, it increased in brightness and magnitude, so that by the time it reached the tops of the trees, the whole wilderness, for some distance around, was illuminated in the most glorious and brilliant manner. **[I] expected to have seen the leaves and the boughs of the trees consumed, as soon as the light came in contact with them**; but, perceiving that it did not produce that effect **[I] was encouraged with the hopes of being able to endure its presence**, which descended gradually

[183] Christensen, Matthew B., The First Vision: A Harmonization of 10 Accounts from the Sacred Grove, Cedar Fort (2014), pp. 27-28.

until it fell upon me. It presently **rested down upon me** and **filled me with unspeakable joy**. [Indeed] it continued descending, slowly, until it rested upon the earth, and [I] was enveloped in the midst of it.

"**I was filled with the Spirit of God**, [and the] Lord opened the heavens upon me. **A personage appeared in the midst of this pillar of flame**, which was spread all around and yet nothing consumed, [having a] light complexion, blue eyes, a piece of white cloth drawn over his shoulders, his right arm b[are]. After a while, another personage soon appeared to the side of the first [who was] like unto the first. I saw two heavenly [and] angelic Personages, who **exactly resembled each other** in features, and likeness, [and] whose brightness and glory defy all description, **standing above me in the air**. One of them spake unto me, calling me by name and said, pointing to the other—*This is My Beloved Son. Hear Him!* He testified also unto me that Jesus Christ is the Son of God."[184]

The Appearance of Moroni to Joseph Smith on September 21, 1823[185]

Some three years later, again in answer to prayer, the boy Prophet received several visitations from an angel who told Joseph that his name was Moroni. Joseph described Moroni's first appearance inside his room in the family cabin in JS-H 1:30-32, 43.

"I discovered a **light appearing** in my room, which **continued to increase until the room was lighter than at noonday**, when **immediately a personage appeared at my bedside**, **standing in the air**, ...

"He had on a loose robe of most exquisite whiteness. It was **a whiteness beyond anything earthly I had ever seen**; nor do I

---

[184] Christensen, Matthew B., The First Vision: A Harmonization of 10 Accounts From the Sacred Grove, Cedar Fort (2014), pp. 27-28

[185] JS-H 1:30-32, 43

believe that any earthly thing could be made to appear **so exceedingly white and brilliant**. ...

"Not only was his robe exceedingly white, but **his whole person was glorious beyond description**, and **his countenance truly like lightning**. **The room was exceedingly light, but not so very bright as immediately around his person**. ...

"After this communication, **I saw the light in the room begin to gather immediately around the person** of him who had been speaking to me, and **it continued to do so until the room was again left dark, except just around him**; when, **instantly I saw, as it were, a conduit open right up into heaven**, and **he ascended till he entirely disappeared**, and the room was left as it had been before this heavenly light had made its appearance." [emphasis added]

Both visitations qualify as miracles since they violate the laws of nature as we understand them. The way light behaves, the speed of travel, and the force of gravity were each violated or overridden. Temporally, light from a point source decreases radially with distance. The formula for the intensity of light is $1/R^2$; this means that as R (the radius or distance from the point source) increases, the intensity decreases. If R increases by 2, the intensity decreases by $1/2^2 = 1/4$. During these visions, light appeared as a pillar and the direction and intensity around the personages could be changed at will. In the case of the First Vision, the Godhead presumably had to come from near Kolob in an instant. Kolob is described as the governing star, probably meaning the largest star with the greatest gravity. The largest stars in our galaxy are near the center of the galaxy, some 25K light-years away. Assuming Kolob is near the center of the Milky Way galaxy, 'in an instant' violates the Special Theory of Relativity which states that nothing can exceed the speed of light. Elohim, Christ, and Moroni were all floating in the air – this violates the General Theory of Relativity which defines how gravity works. The description of these visits of heavenly beings far exceeds any such encounter found in science fiction novels.

Given the foregoing discussion of miracles, the forces of nature, and the control of these forces by the priesthood, does God need a blind, chaotic, unguided, time-consuming, trial and error evolutionary process to create life? The answer appears to be an emphatic 'NO!'

After His Kind, Seed in Itself

On a related topic, the plain and precious scriptures are clear regarding the reproduction of both plants and animals - they reproduce after their kind. No matter how one interprets 'kind,' it seems to rule out major changes in types of life-forms. There are 8 levels of modern biological classification: species, genus, family, order, class, phylum, kingdom, and domain. We will use the letters D, K, P, C, O, F, G, and S to represent these classifications. Michael Behe uses money to compare the classifications.. Eight levels of digits, $DKP,COF.GS, would give something like $876,543.21. The 8 in the $100K column represents domains, and the 1 in the $0.01 column represents species. He then uses Darwin's finches to demonstrate the best that Darwinian evolution can do. "The finch ancestors that first colonized the Galapagos Islands two million years ago are thought to have been related to tanagers, which are in the same family of birds that contain Darwin's finches. ... after two million years of **intense selection**, after eons of the **most fierce evolution**, the ***finch ancestors and descendants differ in only the pennies and dimes column***."[186] [emphasis added] Based upon our previous discussions, the best Darwinian evolution can manage is microevolution – producing changes in species and genus, possibly stretching to the family class. Therefore, **based on biological evidence, 'kind' probably refers to man's classifications of species and genus**.

There are four doctrinal sources regarding the Creation: Genesis 1-3 in the Bible, Moses 2-4, and Abraham 4-5 in the Pearl of Great Price, as well as the temple endowment in The Church of

---

[186] Behe, Darwin Devolves, pp. 152-156

Jesus Christ of Latter-day Saints. Examining Genesis 1, Moses 2, and Abraham 4, verses 11-12, 21, 24-26, we get the table below. Abraham 4:11-12 emphasizes "after his kind" multiple times and in multiple ways.

Scriptural Evidence Against Macroevolution			
Phrase	**Genesis 1**	**Moses 2**	**Abraham 4**
"after his kind"	10X	10X	10X
"seed in itself"	1X	1X	2X
"own likeness" (seeds)	1X	1X	2X
"from its own seed"			3X
"only bring forth the same"			1X
"after our likeness" (Adam & Eve)	1X	1X	1X

TABLE 2. SCRIPTURAL REFERENCES TO 'AFTER HIS KIND' AND 'SEED IN ITSELF'

Abraham 4:11-12

11 And the Gods [plural] said:
Let us prepare the earth to bring forth grass;
the herb yielding seed;
the fruit tree yielding fruit,
after his kind,
whose **seed in itself**
yieldeth its own likeness upon the earth;
and it was so, even as they ordered.
12 And the Gods [plural]
organized the earth to bring forth grass
from its own seed,
and the herb to bring forth herb
from **its own seed**,
yielding seed **after his kind**;
and the earth to bring forth
the tree from **its own seed**, yielding fruit,
whose seed could only bring forth the same
in itself,
after his kind;

and the Gods saw that they were obeyed.
[emphasis added]

Once again, we must decide if the scriptures mean what they say. 'After his kind' or 'seed in itself' or similar phrases are repeated 13X in Genesis, 13X in Moses (the inspired version of Genesis), and 19X in Abraham (another modern scripture). To me, it is decidedly clear that God set up life-forms as reasonably unchangeable creations.

The Church of Jesus Christ of Latter-day Saints Official Statement on Evolution

The Church of Jesus Christ of Latter-day Saints' official stand on the evolution of man is stated in the *Origin of Man*, a 1909 First Presidency Doctrinal Exposition signed by the First Presidency under President Joseph F. Smith. (See Appendix 1 for the full text.) An abbreviated, word-for-word extraction of the *Origin of Man* was reissued in 1925 under President Heber J. Grant and the First Presidency. It was entitled the *Mormon View of Evolution*. See Appendix 1. Key doctrines include the following statements. "**God created man in his own image**, in the image of God created he him; male and female created he them." "He [Jesus Christ] is our elder brother, and we, like him, **are in the image of God**. ***All men and women are in the similitude of the universal Father and Mother*** and are **literally sons and daughters of Deity**. Adam, our great progenitor, '**the first man,**' was, like Christ, **a pre-existent spirit**, and, like Christ, he **took upon him an appropriate body, the body of a man**, and so became a 'living soul.' ***The doctrine of pre-existence pours wonderful flood of light upon the otherwise mysterious problem of man's origin***." "***The Church of Jesus Christ of Latter-day Saints, basing its belief on divine revelation, ancient and modern, proclaims man to be the direct and lineal offspring of Deity***. By his Almighty power God organized the earth, and all that it contains, from spirit and element, which exist co-eternally with himself. **Man is the child of God, formed in the divine image and endowed with divine attributes**, ..."

Some members of the Church state that the Church has no official position on organic evolution. On the other hand, the Church DOES have an official position on Adam and Eve and their descendants, i.e., all of mankind. The doctrine that man is made in the image of God as spirit children is exceptionally clear. That we, like Jesus Christ, also took on physical bodies that are in the image of God and our spirits, is also exceedingly clear. This doctrine is reinforced in the Church's *A Proclamation to the World on The Family*. "**ALL HUMAN BEINGS—male and female—are created in the image of God. Each is a beloved spirit son or daughter of heavenly parents**, and, as such, each has a divine nature and destiny." [187] How can anyone state the Church has no official position on the organic evolution of man?

Recent Quotes by Leaders of the Church of Jesus Christ of Latter-day Saints

Some current Church leaders have made recent comments about evolution though they are not official statements. On March 29, 1987, at a BYU Devotional, Elder Russell M. Nelson[188] gave a talk

---

[187] The Family – A Proclamation to the World, The First Presidency and the Council of the Twelve Apostles of the Church of Jesus Christ of Latter-day Saints, September 23, 19915, Salt Lake City, Utah; https://www.churchofjesuschrist.org/study/manual/the-family-a-proclamation-to-the-world/the-family-a-proclamation-to-the-world?lang=eng

[188] Russell M. Nelson became the President of the Church of Jesus Christ of Latter-day Saints on January 14, 2018. Prior to his calling as an Apostle of Jesus Christ and a member of the Quorum of the Twelve Apostles, he was a world-renown heart surgeon for 29 years. He served as the president of the Society of Vascular Surgery, the Director of the American Board of Thoracic Surgery, and was the president of the Utah Medical Association. He developed one of the first heart-lung bypass machines and was the inventor of the tricuspid valve regurgitation procedure, though he attributes the latter to revelation that came during the surgery itself. https://en.wikipedia.org/wiki/Russell_M._Nelson

entitled *The Magnificence of Man*. Some of his comments follow. "**A marvelous process of genetic coding** is established, by which all the basic human characteristics of the unborn person are determined. ...

"**The eyes with which we see are magnificent**. No doubt you have stood before the mirror, as have I, watching pupils react to changes in the intensity of light—dilating to let more light in, constricting to reduce the light allowed to reach the eyes' sensitive retinas. A self-focusing lens is at the front of each eye. Nerves and muscles synchronize the function of two separate eyes to produce one three-dimensional image. Eyes are connected to the brain, ready to record sights seen. No cords, no batteries, no external connections are needed; **our visual apparatus is marvelous—infinitely more priceless than any camera money can buy**. ...

"**[Some] have deduced that**, because of certain similarities between different forms of life, there has been ***a natural selection of the species, or organic evolution from one form to another***. Many of these people have ***concluded that the universe began as a 'big bang'*** that eventually resulted in the creation of our planet and life upon it. ***To me, such theories are unbelievable!*** **Could an explosion in a printing shop produce a dictionary?** ***It is unthinkable!*** Even if it could be argued to be within a remote realm of possibility, such a dictionary could certainly not heal its own torn pages or renew its own worn corners or reproduce its own subsequent editions! **We are children of God, created by him and formed in his image**.

"***I found that there are at least fifty-five verses of scripture that attest to our divine creation.*** I believe all those scriptures that pertain to the creation of man. But the decision to believe is a spiritual one, not made solely by an understanding of things physical, for we read that "**the natural man receiveth not the things of the Spirit of God: for they are foolishness unto him:**

neither can he know them, because they are spiritually discerned. (1 Cor. 2:14.)"[189] [emphasis added]

Some years later, Elder Russell M. Nelson as a member of the Quorum of the Twelve Apostles, and Elder Wickman, First Counsel of the Seventy and General Counsel for the Church, were interviewed by Robert Ruby, Senior Editor, Pew Forum on Religion & Public Life on May 16, 2007.

"Question: Different denominations deal differently with questions about life's origins and development. Conservative denominations tend to have more trouble with **Darwinian evolution**. **Does the church have an official position on this topic**?

"Elder Nelson: ***We believe that God is our creator and that he has created other forms of life***. It is interesting to me, drawing on *my 40 years' experience as a medical doctor*, how similar those species are. We developed open-heart surgery, for example, experimenting on lower animals simply because the ***same creator made the human being***. We owe a lot to those lower species. But ***to think that man evolved from one species to another is, to me, incomprehensible***.

"Question: Why is that?

"Elder Nelson: ***Man has always been man***. **Dogs have always been dogs. Monkeys have always been monkeys. It's just the way genetics works**."[190] [emphasis added]

The Church of Jesus Christ is concerned about religious liberty in the modern cultural climate. Elder Jeffrey Holland spoke at Chapman University on *Faith, Family, and Religious Freedom* in February 2015.

---

[189] Ensign, Jan. 1988

[190] https://www.pewforum.org/2007/05/16/in-focus-mormonism-in-modern-america/

Speaking of our current cultural environment, Elder Holland expressed **concern about the rise of atheism, specifically mentioning several evolutionists (Dennett and Dawkins)**.

"Our era has been given other labels—post-Christian and post-modern to name two ... Such an age, whatever it is called, has created a climate for popularizing the diminution or minimizing of religious faith in a way that is unprecedented in Western culture, or certainly in American culture. Just so very few years ago anyone openly advocating atheism would surely have had a scarlet 'A' seared upon his or her breast as a warning to all who would come near. But listen now to Richard Dawkins:

"Only the willfully blind could fail to implicate the divisive force of religion in most, if not all, of the violent enmities in the world today. ... ***Those of us who have for years politely concealed our contempt for the dangerous collective delusion of religion need to stand up and speak out.***"[191]

"And many have. After Sam Harris published his provocative The End of Faith in 2004, Christopher Hitchens, Daniel Dennett, Dawkins, himself, and their band of "New Atheists" [referred to as **The Four Horseman of Atheism[192]**] have achieved near-celebrity status publishing a deluge of texts decrying belief in God. Hitchens spoke for most of them when he said, "***One reason I have always detested religion is its sly tendency to insinuate the idea that the universe is designed with 'you' in mind or, even worse, that there***

---

[191] Quoted in the original; Richard Dawkins, *Time to Stand Up*, Freedom from Religion Foundation, September 22, 2001; http://ffrf.org/news/timely-topics/item/14035-time-to-stand-up.

[192] In 2007, Christopher Hitchens, Richard Dawkins, Sam Harris, and Daniel Dennett filmed a landmark discussion about modern atheism; it later appeared in print, The Four Horseman: The Conversation that Sparked the Atheistic Revolution, Random House, 2019; **they were called the Four Horseman of Atheism because "they were the heralds of religion's unraveling**." https://samharris.org/books/the-four-horsemen/

is a divine plan into which one fits."[193] (Of course, Hitchens passed away not long ago and may now have newer views on the idea of a divine plan. And never mind that militant atheism is the ultimate untenable position simply because it would take someone with God's omniscience and omnipresence to be sure that nowhere in the universe was there such an omniscient and omnipresent being - Catch 22.)" This is the same untenable position demonstrated by Korihor when he exclaims "ye cannot know that there shall be a Christ."[194] Alma says emphatically, "what evidence have ye that there is no God, or that Christ cometh not?"[195]

Now that we have an idea of what Intelligent Design is and some relevant justification, Elder Holland's mention of the 'New Atheists' leads us directly into the next section on Anti-Christs. An Anti-Christ is one who teaches and preaches that there is no Christ, that Christ has not come to Earth, and that there is no "great plan of God."[196] Many evolutionary evangelists meet every requirement of being Anti-Christs. This is the topic of the next chapter.

---

[193] Christopher Hitchens, Hitch-22, Atlantic Books, 2010, p. 332

[194] Alma 30:15

[195] Alma 30:40

[196] Alma 34:9

9. ANTI-CHRISTS & EVOLUTION EVANGELISM

The Book of Mormon, a book of scripture preserved for the Saints in the Last Days, details three separate Anti-Christs. The purpose seems to be a warning about Anti-Christs that will occur in our day. The most detailed account of the three is that of Korihor (Alma 30). Korihor, did all of the following: (1) he rejected all religious teachings and the scriptures; (2) he denied the existence of God; (3) he said there was no need for Christ; (4) he said we can know nothing that cannot be seen; (5) he ridiculed any revelations from God; and (6) insisted on a sign. In the end, he confessed that he had been deceived by Satan and taught doctrines pleasing to the carnal mind until he began to believe them himself.

The Creation, the Fall, and the Atonement

There are three key events in the Earth's history: the Gods' Creation, the Fall of Adam and Eve, and Christ's Atonement. The three go together; they are not independent but are intimately linked together. Anti-Christs deny that there are Gods who planned and directed the Creation. Evolutionists deny that there was a Fall; they accept that men evolved from simpler organisms and there was always death. Without the divine Creation of man and the subsequent Fall, there is obviously no need for the Atonement or a Savior. The following is the scriptural account of the Anti-Christ, Korihor, contending with the Lord's prophet, Alma.

Korihor, the Anti-Christ (Alma 30)

The appearance of Korihor preaching

> 6 [Korihor] began to preach … against the prophecies which had been spoken by the prophets, concerning the coming of Christ.
>
> 12 **there should be no Christ**.
>
> 13 **no man can know of anything which is to come**.
>
> 15 **ye cannot know of things which ye do not see**

15 **ye cannot know that there shall be a Christ**.

16 this derangement of your minds ... lead[s] you away into a belief of things which are not so.

17 **every man fared in this life according to the management of the creature**; [survival of the fittest]

17 every man prospered according to his genius, [survival of the fittest]

17 every man conquered according to his strength; [survival of the fittest]

17 whatsoever a man did was no crime. [no absolute truth or morals]

18 **when a man was dead, that was the end thereof**. [evolution]

26 ye do not know that there shall be a Christ.

29 he would revile even against God,

37-38 **Believest thou that there is a God? Nay**

The response of Alma, the prophet, what evidence do you have that there is no God?

40 what evidence have ye that there is no God, or that Christ cometh not?

41 **I have all things as a testimony that these things are true**;

42 I know that thou believest, but ... ye have put off the Spirit of ... the devil has power over you, and he doth carry you about, ... he may destroy the children of God.

43 [Korihor] show me a sign

44 Alma said unto him: Thou hast had signs enough; ... ye have the testimony of all these thy brethren, and also all the holy prophets? The scriptures are laid before thee, yea, and **all things denote there is a God; yea, even the earth, and**

all things that are upon the face of it, yea, and its motion, yea, and also all the planets which move in their regular form do witness that there is a Supreme Creator.

Korihor's response, he requests a sign again:

48 I do not deny the existence of a God,

48 but **I do not believe that there is a God**; and

48 **ye do not know that there is a God**;

48 except ye show me a sign, I will not believe.

Alma's gives Korihor a sign:

49 This will I give unto thee for a sign, that thou shalt be struck dumb, … ye shall no more have utterance.

51 Art thou convinced of the power of God?

Korihor's final response, he admits he was deceived by the devil:

52 I know that nothing save it were the power of God could bring this upon me;

52 yea, and **I always knew that there was a God**.

53 **the devil hath deceived me**; for he appeared unto me in the form of an angel,

53 and said unto me: Go and reclaim this people, for they have all gone astray after an unknown God.

53 And he said unto me: There is no God; yea, and he taught me that which I should say.

53 And I have taught his words; and **I taught them because they were pleasing unto the carnal mind**;

53 I taught them, even until I had much success, [praise of the world]

53 insomuch that **I verily believed that they were true**;

53 and for this cause I withstood the truth.

Given the characteristics of Anti-Christs, it makes sense to see what some evolutionists have to say about Christianity and God. Please compare the list of things that Anti-Christs preach versus the statements that follow from prominent evolutionists and scientists who believe science, naturalism, and evolution have replaced a need for God and religion.

Julian Huxley, an evolutionary biologist, eugenicist, internationalist, and humanist, has been nicknamed Darwin's Bulldog. He was one of the main architects of the modern evolutionary syntheses – Neo Darwinism. "**I believe that [a] drastic reorganization of our pattern of religious thought is now becoming necessary, from a god-centered to an evolutionary-centered pattern**." [197] In other words, God-centered religions should be replaced by a naturalistic, evolution-centered religion.

Richard Dawkins, nicknamed Darwin's Greyhound, made the following statements. "Pretend as they will to scientific credentials, **the anti-evolution propagandists are always religiously motivated**. ... In most cases, they know ... what to believe because their parents recommended an ancient book that tells them what to believe**. [they don't believe evolution because] the holy book of childhood [the Bible] cannot be, must not be, wrong**." [198] [emphasis in the original] "Nearly all peoples have developed their own **creation myth**, and the **Genesis story is just the one that happened to have been adopted by one particular tribe of Middle East herders**."[199] [emphasis added] "**Faith is the great copout, the great excuse to evade the need to think and evaluate evidence**.

---

[197] https://en.wikipedia.org/wiki/Julian_Huxley

[198] Dawkins, The Blind Watchmaker, 1996, p. xi

[199] Dawkins, The Blind Watchmaker, 1996, p. 316

Faith is belief in spite of, even perhaps because of, the lack of evidence."[200]

Compare Dawkins's views to Elder Boyd K. Packer. "As surely as a man may determine whether a purported diamond is genuine by subjecting it to well-known tests for diamonds, **the scriptures may be subjected to well-known tests for scriptures**. "There is a very precise formula. To apply it, one must, of necessity, **move from criticism to spiritual inquiry**. ..._**Many an academic giant is at once a spiritual pygmy and, if so, he is usually a moral weakling as well**. ***Such a man may easily become a self-appointed member of a wrecking crew determined to destroy the works of God.***"[201]

Stephen Hawking, a famous astrophysicist, became more and more enthralled with scientific naturalism as he aged. His last book, The Grand Design, was pointedly atheistic.[202] Lennox wrote a rebuttal, entitled God and Stephen Hawking.[203] Referencing the current inflationary theory of the universe, Hawking states, "**Spontaneous creation** is the reason there is something rather than nothing, ***why the universe exists, why we exist*. It is not necessary to invoke God to ... set the universe going**." While discussing the development of naturalism as a means of understanding the universe, Hawkins states, "The idea arose that nature follows consistent principles that can be deciphered. And so began the long process of **replacing the notion of the reign of the gods** with the concept of the **notion of a universe that is governed by the laws of nature**, and created according to a blueprint we

[200] Richard Dawkins Quotes 2, Canvas Art Print, Amazon.com; https://www.amazon.com/Richard-Dawkins-Quotes-Canvas-Print/dp/B00IIVU6YS/ref=sr_1_2?dchild=1&keywords=quotes+2%2C+canvas+art+print%2C+dawkins&qid=1589328869&sr=8-2

[201] Packer, We Believe All That God Has Revealed, Ensign, April 1974)

[202] Stephen Hawking & Leonard Mlodinow, The Grand Design, 2010

[203] John C. Lennox, God and Stephen Hawking, Lion Hudson, 2011

could someday learn to read." Lennox comments that Hawking deliberately gives the impression "that **the concept of God, or the gods, is a placeholder for human ignorance**."[204] [emphasis added] This is in direct contrast to Isaac Newton, who, when he thought about the universe being governed by natural laws, thought about the single Mind that created it (the blueprint).[205]

Europe has been becoming more secular for decades. The following incident is rather frightening. In June 2007, the European Centre for Law and Justice (ECLJ) presented a brief to the Council of Europe regarding the teaching of creationism in any educational setting outside of religion classes. Reviewers of the brief stated that, "the ECLJ argued that, '[t]he aim of the Report is (1) to forego scientific discussion between the theories of evolution and creationism, or intelligent design, (2) to impede the educational formation of children by restricting classroom exploration of ideas, and (3) effectively infringe on the rights of free exercise of expression, religion, and education. Therefore, this project goes against the respect for pluralism and diversity of ideas and opinions that the Council of Europe claims to stand for.'" The vote by the Council was originally postponed while the ECLJ revised the resolution. Reviewers of the revised resolution stated that the new resolution "**reads more like an alarmist screed against the dangers of creationism and belief** rather than a well-researched and formal pronouncement of a parliamentary body." Continuing, the reviewers' comment, in the resolution "**evolution is not simply a matter of the evolution of humans** and of populations. **Denying it could have serious consequences for the development of our societies**. The **authors** of the resolution do not shy away from revealing why they feel the need to **target creationism and brands its supporters as religious extremists** ..." Granted, the EU does not

---

[204] Lennox, God and Stephen Hawking, pp. 16, 23

[205] Nancy R. Pearcey and Charles B. Thaxton, <u>The Soul of Science: Christian Faith and Natural Philosophy, Vol. 16</u>, Crossway Books, 1994

have a Bill of Rights. Nevertheless, the second-most democratic republic in the world has specifically targeted religious expression.[206]

The United States of America is itself, not free of anti-religious bias. The National Center for Science Education (NCSE) is a taxpayer-funded organization. Their stated mission is to "provide information and advice as the premier institution **dedicated to keeping evolution in the science classroom and creationism out.**"[207] On their website is an article entitled *What is 'Intelligent Design' Creationism?*[208] It is found just above a "**Donate Now: If you support accurate science education, support NCSE**," i.e., if you do not support creationism or intelligent design.[209] [emphasis added] The article explains that Intelligent Design (ID) is a 1980's successor to 'creation science' which began in the 1960s. Then in the 1990s, ID advocates encouraged the teaching of ID as an alternative to evolution. The article states that in the 2000s, ID advocates began to "encourage teachers to teach students wrongly that there is a 'controversy' among scientists over whether evolution has occurred." The NCSE has testified at most if not all State School Board discussions nationwide about whether to allow the teaching of creationism or intelligent design along with Darwinian evolution in public schools. The director, Eugenie Scott, is a rabid 'defender of the faith.'[210]

Carl Sagan, a noted astronomer, firm evolution proponent, and science popularizer argued that "Science is more than a body of knowledge; it is a way of thinking." ... "**The Religious Right, realize**

---

[206] https://eclj.org/COE/council-of-europe-passes-resolution-to-ban-creationism-from-classroom

[207] NCSE website, December 2008

[208] October 17, 2008

[209] https://ncse.ngo/what-intelligent-design-creationism

[210] https://ncse.ngo/

this, and they see it as a way of thinking inimical to their dogmatic assertions, assertions drawn on a Bronze Age religion codified in a document put to paper in the first millennium B.C.E."[211] One must wonder what his thoughts are now that he is in the Spirit World?

Charles Shermer is the founder of The Skeptics Society and is the editor-in-chief of the Skeptic magazine. In the not-to-distance past, he wrote a column for the Scientific American magazine entitled *The Skeptic* for over a decade. Shermer proudly displays a large bust of Darwin in his office. He stated the following in several of his books arguing against God, intelligent design, pseudoscience, and religion. "Darwin matters because evolution matters. **Evolution matters because science matters**. Science matters because it is the preeminent story of our age, **an epic saga about who we are, where we came from and where we are going**."[212] "There is a significant difference between having no belief in a God and believing there is no God," and "**the whole point of faith, in fact, is to believe regardless of the evidence, which is the very antithesis of science**."[213] According to Shermer, we are animals governed by instinct having no special significance among life-forms, **we descended from apes**, and **when we die that is the "end thereof."** This is identical to what Korihor, the Anti-Christ, said to Alma.[214]

Which do you believe, Shermer's explanation of where we came from, who we are, and where we are going or the Lord's

---

[211] Sagan, The Demon-haunted World: Science As a Candle in the Dark, Random House, 1996

[212] Michael Shermer, Why Darwin Matters: The Case Against Intelligent Design, Macmillan, 2007, p.161

[213] Michael Shermer, How We Believe: Science, Skepticism, and the Search for God, Macmillan, 2003, p.271

[214] Alma 30:18

explanation? In the Lord's plan, the creation of the universe and the earth was to bring about the "immortality and eternal life of man."[215] Man is an everlasting intelligence, embodied with a spirit body by our Heavenly Father and Heavenly Mother. We were then born to mortal parents to gain a physical body and "learn by our own experience the good from the evil." After we die, our spirits return to the Spirit World awaiting the resurrection (the reuniting of spirit and physical body) enabled by the Atonement of Jesus Christ. Through our own experience, by our endeavor to become more Christ-like, and through our repentance (also enabled by the Atonement), we may return to the Celestial Kingdom where our heavenly parents reside and continue our eternal existence.

Zimmer, who wrote the book, Evolution[216] as a companion volume to the 7-hour PBS television documentary,[217] wrote the following provocative statements.

"**Evolution substituted a naturalistic explanation of cold comfort for our former conviction that a benevolent deity fashioned us directly in his own image**, to have dominion over the entire earth ..."

"**Darwinian natural selection offers no solace or support for these traditional hopes about human necessity or cosmic importance**."

Darwin's theory has "far-reaching and radical philosophical consequences" ... it ***removes "any notion of an inherent goal or purpose***."

"***Only if God had directly created mankind in His image was there any basis for morality. ... If we were the product of nothing***

---

[215] Moses 1:39

[216] Zimmer, Evolution, 2006, pp. xi, xii, xii, 315

[217] Evolution, a 7-hour PBS television documentary, September 24-27, 2001

more than natural selection, how could we be God's special creation? And if we were not, then why should people adhere to the Bible?"

Wow! Zimmer is not a scientist. He is a writer. He had many evolutionists assisting him as advisors. It seems that without their names directly attached to the above quotes, they felt free to express their true animosity towards God, religion, and even morality.

Dr. Axe states that the most significant cost of teaching evolution in grade school and college is the "systematic devaluation of human life." Continuing, he says "contemplate for just a moment the dystopian vision of a generation of human beings believing in their hearts that they are nothing more than bestial accidents fending for themselves in a world where morality is fiction, and you begin to grasp the true stakes."[218]

Icons of Evolution

While completing his Ph.D. in cell and developmental biology at the University of California, Berkeley, Jonathan Wells noticed that "all of my textbooks dealing with evolutionary biology contained a blatant misrepresentation." Ernst Haeckel, in 1891, drew several vertebrate embryos side-by-side to show their similarities. The purpose was to provide evidence for common descent. Wells believes that the drawings were distorted and images of earlier embryo stages, which do not look similar, were omitted. There has been much debate about the images since then, with evolutionists claiming there was no intent to deceive (once again, NCSE comes to the evolutionists' defense).[219] Yet the images still ubiquitously occur without explanation. The intent seems to be to deceive readers into believing the evidence for a common

---

[218] Douglas Axe, Undeniable: How Biology Confirms Our Intuition That Life Is Designed, Harper One (2016), pp. 54-55

[219] https://ncse.ngo/accuracy-embryo-illustrations

ancestor. Wells says that "I discovered other biologists have noticed most of them and have even criticized them in print. But their **criticisms have been ignored**. The **pattern is consistent and suggests more than simple error**. At the very least, it suggests that Darwinism encourages distortions of the truth. ... the result is clear: students and the public are being systematically misinformed about the evidence for evolution."[220] In his book, the Icons of Evolution, Wells discusses ten icons in detail. We have discussed five of these: the Miller-Urey experiment (origin of life), Darwin's tree of life (Cambrian explosion), Darwin's finches and fruit flies (both in how evolution really works), and Haekel's embryos (immediately above). The other five are homology, *archaeopteryx*, peppered moths, fossil horses, and man descending from apes (see Appendix 2).

If evolution is 'true,' and the evidence is so overwhelming, why the need to deceive? Satan is a master deceiver. "Satan, ... is the father of all lies, to deceive and to blind men ..."[221] When speaking of the Last Days, the Lord states that they who "have taken the Holy Spirit for their guide, and have not been deceived ... shall abide the day."[222] This states emphatically that in the Last Days, men and women will need the Holy Ghost to keep from being deceived. A great example of this follows.

In the November 1999 issue of National Geographic, an article was published with much fanfare. A fossil had been discovered that was "a missing link between terrestrial dinosaurs and birds that could actually fly." It was named *archaeoraptor* and had come from China. Evolutionists are always looking for missing links – since they are extremely rare. Unfortunately for evolutionists, it was discovered that the fossil was a composite of several fossils. Those

---

[220] Wells, Icons of Evolution, 2000, xii

[221] Moses 4:4

[222] D&C 45:57

involved in selling fossils knew what would sell! On February 3, 2000, National Geographic issued a press release confirming the finding. They immediately began an investigation. In October 2000, National Geographic published the results of their investigation which said that virtually everyone involved in the project had made some mistakes.[223] One must ask, why? The obvious reason appears to have been their eagerness to publish anything that provides evidence for evolution. This was embarrassing to such a prestigious organization. To their credit, they promptly and openly admitted their mistake. The event, however, **highlights a weakness among evolutionists: their desperate eagerness to grasp at anything buttressing their cause and worldview.**

The Department of Biological Sciences, Lehigh University

In the acknowledgments in Michael Behe's book, Darwin Devolves, Behe quietly states, "I am also indebted to several reviewers who did not wish to be mentioned by name." I have noticed this in several books promoting intelligent design. These reviewers did not want their names associated with Behe or ID in any way. This is borne out by a very unusual posting on Lehigh University's Biological Sciences webpage where Dr. Behe teaches biochemistry.[224] It is worth quoting the Department's position on evolution and 'intelligent design' in its entirety ...

"The faculty in the Department of Biological Sciences is committed to the highest standards of scientific integrity and academic function. This commitment carries with it **unwavering support for academic freedom** and **the free exchange of ideas**. It also demands **the utmost respect for the scientific method**, integrity in the conduct of research, and recognition that the validity of any scientific model comes only as a result of **rational**

---

[223] https://en.wikipedia.org/wiki/Archaeoraptor

[224] https://www.lehigh.edu/~inbios/News/evolution.html

hypothesis testing, sound experimentation, and findings that can be replicated by others.

"The department faculty, then, are unequivocal in their support of evolutionary theory, which has its roots in the seminal work of Charles Darwin and has been supported by findings accumulated over 140 years. The **sole dissenter** from this position, Prof. Michael Behe, is a well-known proponent of "intelligent design." **While we respect Prof. Behe's right to express his views, they are his alone and are in no way endorsed by the department.** ***It is our collective position that intelligent design has no basis in science, has not been tested experimentally, and should not be regarded as scientific."*** I know of nothing comparable! This is quite a tribute to Behe and his arguments against evolution. It is also an indictment against the ivory towers of academia.

Now the last quote from Gelernter. Gelernter reluctantly came to abandon evolutionary theory after reading several ID books. He comes to the fray recently, from outside the contentious evolutionary debate – Gelernter is a Computer Scientist. His observation about how some 'ID-haters' behave is spot on.

"Some ID-haters have shown themselves **willing to use any argument – fair or not, true or not, ad hominem or not – to keep this dangerous idea locked in a box forever**. They remind us of the extent to which ***Darwinism is no longer just a scientific theory but the basis of WORLDVIEW, and an emergency replacement RELIGION for the many troubled souls who need one***." [225] [emphasis added]

In summary, ***the fight is about worldviews, not evolution versus intelligent design***. To quote John Lennox again, the "real

[225] Gelernter, *Giving Up Darwin*, Review of Books, Claremont

conflict … is not science vs. religion. It is ***theism versus atheism***, …"[226]

The next section discusses some thoughts on the pre-Adam and pre-Eve creation of the Earth. Some details are accepted as scripture, such as God residing on a planet near a massive star, Kolob. Other details are statements concerning the creation from some of the early Saints who heard the Prophet Joseph Smith speak about it. Some modern church leaders are also quoted. Why is this section pertinent to our discussion about evolution? One, the details are in harmony with the fact that the Earth and man were created purposefully. Second, some of the details completely invalidate some scientific assumptions about such things as the age of the Earth.

---

[226] Lennox, God and Stephen Hawking, 2011, p. 12

10. THOUGHTS ABOUT THE CREATION OF THE EARTH BEFORE ADAM AND EVE

Why discuss the pre-Adam and pre-Eve creation of the Earth? If one is Christian, they presumably believe in the Fall of Adam and Eve. The Fall presupposes the Garden. The Garden was part of the Creation summarized in Genesis, the Book of Moses, the Book of Abraham, and the temple endowment. The obvious questions, then, are where did the Earth come from and where did Adam and Eve come from? Evolutionists would say that the earth was a random assembly of cosmic dust created entirely by natural forces with no involvement of an Intelligent Designer. Additionally, they would say that Adam and Eve were random organisms that came about through millions of years of blind, trial and error, chance-based evolution. Proponents of ID would say that God had a lot to do with the creation of the Earth and everything to do with the creations of Adam and Eve. Which is it? This chapter will discuss the Gods' involvement in the Creation of the Earth, the Fall of Adam and Eve, and Christ's Atonement as part of an overall Plan of Salvation established by God, The Intelligent Designer.

This section will discuss two topics: (1) the spirit creation that preceded the Creation of the Earth; and (2) the physical Creation of the Earth. The Pre-Creation section is about the Plan of Salvation which involves intelligences, spirit bodies, and physical bodies. It will tie the Creation, the Fall, and the Atonement together. It will also discuss the eternal progression of intelligences as they gain, first, spirit bodies and, second, physical bodies. The Creation of the Earth section will discuss the possible location of the creation of the Earth, what probably happened to the Earth after the Fall, and the Old-Earth versus the New-Earth points of view. Through the scriptures, we know the WHY of the Creation. The scriptures, however, provide limited information about the HOW or WHEN of the Creation. We do know some details of the how and when from the scriptures and have some additional hints from the writings of early Church leaders acquainted with the Prophet Joseph Smith.

The Pre-Creation of the Earth

The pre-mortal spirit creation of the Earth, plants, animals, and Adam and Eve are all dictated by doctrine. The key doctrine is the Plan of Salvation, also called the Great Plan of Happiness, the Plan of Mercy, and the Plan of Exaltation. The principle elements of the Plan of Salvation encompass the Creation, the Fall of Adam and Eve, and the Atonement of Jesus Christ – these indispensable events are inextricably interwoven together. The doctrines of eternal, pre-existent intelligences, the need for spirit bodies, and the need for physical bodies are also essential elements of the Plan of Salvation. Each of these doctrines will be discussed.

The Plan of Exaltation

God's plan provided for pre-existing intelligences to advance by acquiring a spirit body through birth. While God's spirit children were in the pre-existence, Father-in-Heaven called the Council in Heaven. Father-in-Heaven, Mother-in-Heaven, Jesus Christ, Michael (Adam), Eve, and others discussed the Plan in every detail. The Plan was that God's spirit children would come to earth through human birth to gain a physical body, would choose to be obedient or not, would experience opposition and trials, and would learn by their own choices the good from the evil. Since we would all make mistakes, Lucifer proposed to force us to follow his dictates by surrendering our moral agency. This was unacceptable to the Gods. Consequently, Christ cast Lucifer out. Unfortunately, a "third part"[227] of God's spirit children followed Lucifer and were denied the opportunity to gain a physical body and progress. Christ volunteered himself to be Our Savior to take upon himself our sins and imperfections on condition of our repentance. He also enabled

[227] D&C 29:36: Rev 12:4; JST Rev 12:4

each of us to be resurrected and unconditionally saved from physical death.[228]

Intelligences

All things were at first eternal, self-existing intelligences. "Man was also in the beginning with God. **Intelligence**, or the light of truth, was **not created or made**, neither indeed can be." [229] [emphasis added] Intelligences have always existed. "**God**, himself, finding he was in the midst of spirits and glory, because he was more intelligent, **saw proper to institute laws whereby the rest could have a privilege to advance like himself**."[230] Key enabling provisions of God's laws were moral agency (to make choices) and opposition in all things (choices). "It must needs be, that there is an opposition in all things. If not so, … righteousness could not be brought to pass, neither wickedness."[231] The scriptures state that each intelligence must "act for itself, … otherwise, there is no existence."[232] The very definition of existence encompasses the ability to act, to make choices. Through our experiences and our choices, man must learn the good from the evil. Man, and probably animals, act. Plants may or may not be capable of acting. Inanimate matter is acted upon. God's purpose in all this is well known, "**This is my work and my glory - to bring to pass the immortality and eternal life of man**." [233] [emphasis added] In summary, the

---

[228] The Church of Jesus Christ.org website, Gospel Topics: Plan of Salvation; https://www.churchofjesuschrist.org/study/manual/gospel-topics/plan-of-salvation?lang=eng

[229] D&C 93:29

[230] Compiled by Joseph Fielding Smith, Teachings of the Prophet Joseph Smith, Deseret Book, 1974, p. 354

[231] 2 Nephi 2:11

[232] D&C 93:30

[233] Moses 1:39

scriptures emphasize that "**there is a God**, and **he hath created all things**, both the **heavens and the earth, and all things that in them are**, both **things to act** and **things to be acted upon**."[234] [emphasis added]

Spirit Bodies

The next step was to embody intelligences with spirit bodies composed of spirit matter. Before coming to Earth, all men and women were the literal spirit offspring of a Heavenly Father and a Heavenly Mother. "All human beings – male and female – are **created in the image of God**. Each is a **beloved spirit son or daughter of heavenly parents**."[235] Other lesser intelligences like animals, plants, and matter were similarly provided with their spirit bodies. The Prophet Joseph Smith recorded the following revelation, "There is no such thing as immaterial matter. **All spirit is matter**, but it is more fine or pure, and can only be discerned by purer eyes; We cannot see it; but when our bodies are purified, we shall see that it is all matter."[236]

Spirit bodies must have added capabilities enabling progression that intelligences alone did not have. Consequently, obtaining spirit bodies was a necessary part of our eternal progression. Spirit bodies must certainly have had amplified perceptions, sensations, appetites, and passions. In the spirit world, the Gods (Father-in-Heaven, Mother-in-Heaven, Jesus Christ, the Holy Ghost, and possibly others) presented their plan to Create an Earth and to provide the opportunity for each of us to gain a physical body by going to the Earth to live. On the Earth, in physical bodies, we would continue to learn the good from the evil at an accelerated rate and to be tried and tested – to see if we

---

[234] 2 Nephi 2:14

[235] The Church of Jesus Christ of Latter-day Saints, *The Family – A Proclamation to the World*, 1995

[236] D&C 131:7-8

would learn to control our newfound physical bodies by our spirits. **Like spirit bodies, our physical bodies provided additional capabilities that our spirit bodies alone did not have**. Joseph Smith said, "We came to Earth that we might gain a body ...**the great principle of happiness consists in having a body**."[237] President Joseph F. Smith had a vision of the spirit world where spirits go after the death of their physical bodies. These spirits "looked upon the long absence of their spirits from their bodies as a bondage."[238] Except for Adam, Eve, and Jesus Christ, all human beings have obtained a physical body through the process of procreation and birth by mortal parents. For Adam and Eve, embodying their spirits with physical matter most likely began in the Garden of Eden.

Physical Bodies

When placed in the Garden, Adam and Eve were immortal. Not glorified but immortal. Adam and Eve could not improve until they became mortal – more on that momentarily. Now we know that Jesus' father was Father-in-Heaven and his mother was Mary. From his glorified and celestialized father, Christ's physical body had the power to live forever (he was immortal). From his mortal mother, his physical body had the power to die (he was also mortal). When Christ died, He chose to allow his spirit to leave His physical body as part of the Atonement. Obviously then, God is capable of procreating physical children, not just spirit children. Normally, the Father and the Mother delegate the sacred privilege of creating mortal bodies for their spirit children to earthly parents.

So, where did Adam and Eve come from? There are several possibilities: they evolved on the Earth, they came from another planet, or they were created here. This book, common sense, and the Doctrinal Exposition *the Origin of Man* render the idea that Adam and Eve evolved unnecessary and contrary to the Plan of

---

[237] Smith, The Teachings of the Prophet Joseph Smith, p. 181

[238] D&C 138:50

Salvation. If they came from another planet, how were they created? Were they born? Who were their parents? I would suggest they were created here on Earth, as they were the dust of this Earth. I also suggest, that like Christ, God the Father was their Father. I would also suggest that God the Mother was their Mother. They were, therefore, immortal like Christ but not glorified. Consequently, Adam and Eve had to Fall to become mortal.

Adam and Eve were born of Father and Mother-in-Heaven with immortal physical bodies – they could not die. God cannot create corruption. The following scriptures and quotes strongly suggest this being the case. Quotes from a few of the early twelve apostles and a few of the early Saints who were acquainted with the Prophet Joseph Smith are of particular importance. They heard from the Prophet's own mouth his teachings on the Plan of Exaltation, the nature of God, his visions, his reasonings, and his insights into man and the creation. Many of these experiences occurred during the Kirtland and Nauvoo eras when the Prophet was working on reading and translating the papyri of Abraham and Joseph. While not revered as scripture, their statements are insightful.

Joseph Smith is reported to have said "God was the great head of human procreation – [God] was truly the father of both our spirits and our [physical] bodies."[239] Regarding Adam's genealogy, the New Testament states "Enos, ... was the son of Seth, which was the son of Adam, which was the son of God."[240] The Book of Moses states, "This is the genealogy of the sons of Adam, who was the son of God, with whom God, himself, conversed." [241] Further, the scriptures say that Adam "became of dust a living soul"[242] and "the

---

[239] Andrus, God, Man, and the Universe, p. 351

[240] Luke 3:38

[241] Moses 6:22; Note that Moses is the JST or inspired version of Genesis.

[242] Moses 6:59

Gods formed man from the dust of the ground, and took his spirit (that is, the man's spirit) and put it into him."[243] This could mean that our Heavenly Parents partook of the food in the Garden which then became part of their bodies. Consequently, when Adam and Eve were born, they were literally made of the dust (the elements) of this Earth.

Other early Church leaders commented on Adam's and Eve's physical births. Heber C. Kimball commented that through the process of procreation, God first "brought forth spirits" in the pre-earth state, after which "he commenced and brought forth tabernacles for those spirits to dwell in." Elder Kimball added, "I came through him, both spirit and body."[244] President Brigham Young declared that Elohim has the power by procreation "to produce both spirit and body."[245] In the scriptures, Elohim is God the Father and Jesus Christ is Jehovah. Previously, President Young said, ***"He created man, as we create our children; for there is no other process of creation in heaven, on the earth, in the earth, or under the earth, or in all the eternities, that is, that were, or that ever will be."***[246] Finally, Joseph F. Smith explained, "Adam, our earthly parent, was also born of woman into this world, the same as Jesus and you and I."[247]

It is logical to believe that a similar procreation process was used to populate the Earth with animal life. Presumably, celestial animals were brought here after which their offspring were immortal as were Adam and Eve. For plants, seeds were brought from other worlds since the Gods "planted a garden eastward in

---

[243] Abraham 5:7; also, Moses 3:7

[244] JD 6:31

[245] JD 15:137

[246] JD 11:121

[247] Deseret Evening News, December 27, 1913, Sec. III, p. 7

Eden."[248] Brigham Young said that "the seeds of every plant composing the vegetable kingdom were brought from another world."[249] Hyrum Andrus reiterates that the ***procreation of man and animals and the transplantation of seeds are the keys to the creation process of living organisms***.[250]

The Creation of the Earth

There are four extant accounts of the Creation: Genesis, Moses, Abraham, and temple endowment of The Church of Jesus Christ of Latter-day Saints. Moses is an inspired translation of the Book of Genesis by the Prophet Joseph Smith to restore many plain and precious things that had been lost. Abraham is an account of some of Abraham's and Sarah's journeys, especially their journey into Egypt. It also contains a marvelous vision that Abraham had of the creation. According to Rodney Turner, Moses viewed the creation "from the standpoint of this planet in its finished and fallen state. He figuratively looked **backward**." Abraham's "unique vantage point [was] from the standpoint of eternity. He looked **forward** in time to the yet-to-be-done work of the Gods."[251] [emphasis in the original]

The temple account includes the same elements as the scriptures with some order changes. The general order of the Creation makes logical sense. Mankind and some animals require both plant and animal nourishment while in a mortal state. Other animals require nourishment from plants. Consequently, the later order of the Creation must be plants preceding animals, and animals preceding humans. Plants require earth, air, water, and sunlight. Air and water require something to rest upon, consequently, the earth must have come first. A likely order of

---

[248] Abr 5:8; Moses 3:8

[249] JD 7:285-6

[250] Andrus, <u>God, Man, and the Universe</u>, p. 360

[251] Turner, <u>The Footstool of God</u>, pp. 80-81

creation is the **earth**, then **water and the atmosphere**, followed by **light**, then **plants**, **animals**, and **humankind**. The scriptures speak of the Creation periods as 'days' or 'times.'

On day number four, according to Genesis, Moses, and Abraham, the lights which were destined to shine on the Earth were made ready.[252] However, the light that illuminated the newly created Earth was Kolob, not the sun nor the moon. Before Abraham saw the Earth, the Lord showed him the order of the stars. Larger stars 'governed' smaller stars and revolved [and rotated] more slowly. "And the Lord said unto me: These are the governing ones; and the name of the great one is Kolob, because it is near unto me. ... Kolob [is] after the manner of the Lord, according to its times and seasons in the revolutions [and rotations] thereof; that one revolution [rotation] was a day unto the Lord, after his manner of reckoning, it being one thousand years according to the time appointed unto that whereon thou standest."[253] Abraham then saw Adam and Eve in the Garden of Eden and "saw that it was after the Lord's time, which was after the time of Kolob; for as yet the Gods had not appointed unto Adam his reckoning."[254]

The description of Fig. 1 in Facsimile No. 2 in the Book of Abraham is remarkably interesting. "Kolob, signifying the first creation, nearest to the celestial, or the residence of God. First in government, the last pertaining to the measurement of time. The measurement according to celestial time, which celestial time signifies **one day to a cubit**. **One day in Kolob is equal to a**

---

[252] Gen. 1:14-19; Moses 2:14-19; Abra. 4:14-19

[253] Abr. 3:3-4, 9-10, 16; Note: the word revolve can mean both revolving around another object as well as rotating on its own axis. I will use revolve to mean orbiting and rotate to mean rotating on its own axis. See also 2 Peter 3:8.

[254] Abr. 5:13

thousand years according to the measurement of this earth."[255] Again, Abraham is instructed that one day on Kolob is equal to 1000 years on the Earth, but then the Lord states that there is "one [celestial] day to a cubit." Here the Lord is giving Abraham some measuring tools – time and distance according to the Lord and Abraham's vision of the heavens. Because of the vastness of space, modern astronomers have invented special measures of distance such as light-years and AU's (astronomical units). A light-year is the distance light travels in one year, equal to ~9.5×10^{12} km or 5.9×10^{12} miles. One AU is the distance from the Earth to the Sun. The current distance from the Earth to the Sun is 150M km or 93M miles. In 1000 years, the Earth would revolve around the Sun 1000X. The total distance traveled would be ~600 B miles.[256] The distance light travels in one year, is ~6,000 B miles (~6T miles). Consequently, Abraham's Kolob-cubit is ~1/10 of a light-year.

As already noted, the Earth was created near Kolob, not near its destination in our solar system. John Taylor stated that **the earth "was first organized near the planet Kolob."**[257] Likewise, Brigham Young remarked that **"when the earth was framed ..., it was near the throne of our Father in Heaven."**[258] The fact that the Earth was created near Kolob does not correlate with current astronomical theories that the Earth condensed from dust and particles surrounding the Sun as did all the other planets and moons in our solar system. The solar system was purposely created in preparation for the arrival of the mortal earth. The solar system, itself, may have condensed from space dust under the Lord's

---

[255] Facsimile No. 2, Fig. 1, Book of Abraham

[256] Clifford James Stratton, The Facsimiles in the Pearl of Great Price: The Prophet Joseph Smith's 'Explanations, self-published

[257] John Taylor, *Origin and Destiny of Woman*, The Mormon, New York, August 29, 1857

[258] JD 17:143

direction. Newton himself "believed that the solar system could not be the product of random natural causes that were merely … 'blind and fortuitous.' In his General Scholium, Newton argued that the solar system had to be the product of a Cause that was 'very well skilled in mechanics and geometry.' He wrote: 'This most beautiful system of sun, planets, and comets could only proceed from the counsel and dominion of an intelligent and powerful Being.'"[259]

The earth abides the celestial law.[260] However, it is destined to pass through several states of glory. Before the Fall of Adam, the Earth was immortal as was Adam. Adam, Eve, and the Earth were in a Terrestrial state. After the Fall, the Earth became mortal – it became a Telestial abode. When our Savior comes again, at the beginning of the Millennium, the Earth will return to a Terrestrial state. After every living organism on the Earth is resurrected, the Earth will die and be resurrected to its final Celestial glory. "This death, … is a penalty of the original sin."[261] "What is it that will make the earth die? It will be the withdrawing of the spiritual portion from it, that which gives it life."[262] "The resurrection of all life-forms will require that their spirits leave … before the earth's own spirit and body are separated by death."[263] Note that the Earth is a living entity. When the Earth is resurrected, it will become the abode of her celestial inhabitants and the Earth will return to its birthplace near Kolob. The Prophet Joseph Smith prophesied that **"this earth will be rolled back into the presence of God, and [be]**

---

[259] Tina Simmons, *Isaac Newton in Quarantine*, American Thinker, May 2, 2020

[260] D&C 88:20

[261] JD 1:331

[262] JD 21:201

[263] Turner, The Footstool of the Lord, p. 223

crowned with celestial glory."[264] (TPJS, p. 181) Brigham Young taught that **this earth "will go back into the presence of God, where it was framed."**[265]

The Garden and the Fall

While the Earth orbited Kolob, Adam and Eve were placed in the Garden. Even though the Creation of the Earth and the Garden were physical, Adam, Eve, the Earth, and all its inhabitants were immortal. Descendants of celestial parentage cannot be corrupt. In addition, the consequences of becoming mortal were so great, it had to be Adam's and Eve's decisions to enter mortality. "I, the Lord God, commanded the man, saying: Of every tree of the garden thou mayest freely eat, but of the tree of the knowledge of good and evil, thou shalt not eat of it, nevertheless, **thou mayest choose for thyself, for it is given unto thee**; but, remember that I forbid it, for in the day thou eatest thereof thou shalt surely die."[266] Therefore, the Fall introduced both physical death (from eating the fruit) and spiritual death (from disobeying a commandment).

When Adam and Eve partook of the forbidden fruit, they became fallen creatures. The rest of the Earth's inhabitants followed suit as did the Earth itself. At this point in time, the Earth fell through space from its orbit near Kolob to its current location in our solar system. Brigham Young said, **"when man fell, the earth fell into space and took up its abode in this planetary system, and the sun became our light."**[267] John Taylor explained that **the earth "had fled and fallen from where it was organized near the planet Kolob."**[268] With the Fall, man's time reckoning also changed from

---

[264] Smith, Teachings of the Prophet Joseph Smith, p. 181

[265] JD 9:317; 7:163, 8:8

[266] Moses 3:16-17

[267] JD 17:143 (Brigham Young)

[268] John Taylor, The Mormon, August 29, 1857

the Lord's time to our current time reckoning of days, seasons, and years.

The Fall of man necessitated the infinite Atonement of a God, Jesus Christ. Immortality comes as an unconditional gift from Jesus Christ, our Savior, and our Redeemer, through his resurrection and His conquering physical death. "For as in Adam all die, even so in Christ shall all be made alive."[269] Eternal life is a conditional gift. Individuals must qualify themselves for eternal life by keeping God's commandments, repenting, forgiving others, and enduring to the end. Eternal life is life in the Celestial Kingdom, the same life that Heavenly Father and Heavenly Mother have – exemplified in the perpetuation of the family. "***THE FAMILY is ordained of God. Marriage between man and woman is essential to His eternal plan***."[270] The Creation was purposeful; designed for men to gain a physical body, to learn to distinguish the good from the evil, to learn to choose the good, to repent after choosing evil, and to begin a family.

The doctrine of the Plan of Salvation puts the Creation in an eternal perspective. Consequently, modern science's assumptions about the formation of the universe, the solar system, the earth, and mankind are overruled. Nevertheless, the Lord has left open many questions. Sometime in the future, the answers to all these questions will be given. "Yea, verily I say unto you, in that day when the Lord shall come, he shall reveal all things—Things which have passed, and hidden things which no man knew, **things of the earth, by which it was made, and the purpose and the end thereof**."[271]

---

[269] 1 Cor. 15:22

[270] *The Family – A Proclamation to the World*, The First Presidency and the Council of the Twelve Apostles of the Church of Jesus Christ of Latter-day Saints, 1995

[271] D&C 101:32-33

How Long Did the Pre-Garden Creation Take?

One obvious question is how much time did the Pre-Garden Creation take? There are two fundamental options: a **Young Earth** (~13K years old) or an **Old-Earth** (~4.6 B years old). From a scriptural standpoint, one could take a 'day' in Genesis to mean 1000 years.[272] In this case, the Creation and the Fall took 7000 years (7 days). After the Fall, the Earth was destined to live another 7000 years. D&C 77:6-7 tells us that the seven seals spoken of in revelation each represent about 1000 years, meaning 7000 years of this "earth's ... temporal existence."[273] We are now ending the 6th Seal, meaning 6000 years since the Fall (6 more days). Consequently, the earth is now ~13,000 years old.

If one takes 'day' in Genesis, Moses, and Abraham and 'time' in Abraham to be a period, these periods could be different, undetermined lengths of time. According to modern science, the Earth is ~4.6 B years old. According to a letter from W. W. Phelps in the Church's periodical, *The Times and Seasons*, the mortal Earth is 2.56 B years old. This number can be arrived at by taking 7000 years X 365 days/year X 1000 years/day. Using 13K years for the entire age of the earth, the age of the earth is 4.7 B years old; only 2% off modern science's estimate! Was this number calculated by Phelps or the Prophet Joseph Smith? We do not know. However, it raises the possibility that early members of the Church had discussions of an Old-Earth.

Contemporary to Darwin, Charles Lyell published Principles of Geology in three volumes (1830-1833). Darwin took Volume 1 with him on his famous voyage. Lyell's position was that the present is the key to the past and uniformitarianism reigned – things happen slowly and continuously and are the same now as they were in the past. What science has discovered, however, is that the earth and

---

[272] JST 2 Peter 3:8

[273] D&C 77:6-7

evolution are anything but uniform – catastrophism is the norm. Scientists estimate the number of mass extinctions during the last 540M years (from the end of the Cambrian to the present) at 5 to more than 20.[274] Additionally, it is also evident that geographic features can change rather dramatically, almost instantaneously. When Mt. St. Helens erupted, much of the snow and ice on one side of the mountain instantly melted. The water, snow, and ice rushed down the mountain forming deep gorges (150 feet deep) for miles and some areas were covered with up to 600 feet deep of mud and ash up to 17 miles away – in a few hours!

Church leaders have varied opinions on the age of the Earth. Those advocating a Young Earth include President Joseph Fielding Smith, Elder Bruce R. McConkie, Hyrum L. Andrus, Rodney Turner, Melvin A. Cook, and M. Garfield Cook. Those advocating an Old-Earth include Elder James E. Talmage, Elder John A. Widtsoe, B. H. Roberts, J. Ward Moody, and Michael J. Rhodes. There are theologians, scientists, and engineers on both sides of the debate.

Another question is how are the dates of geologic layers determined? Geological layers are called strata. In general, rocks and fossils in a column (a vertical slice) of strata are assumed to be older than those on top of it and younger than those below it. Fossils are often used to date strata. Alternately, fossil ages may be determined by what strata they occurred in – this can result in problematic circular arguments. Other methods can sometimes be used, such as examining the direction of magnetization in magnetic materials (the earth's magnetic north is constantly moving at a presumed fixed-rate and known direction) or by radiometric dating.

A good example of radiometric dating is carbon-14 dating (C-14). The dominant carbon isotope is carbon-12 which has 6 protons and 6 neutrons in its nucleus. Carbon-14, on the other hand, is a

---

274 https://en.wikipedia.org/wiki/Extinction_event

trace isotope having 6 protons and 8 neutrons in its nucleus. About 1-2 atoms out of every 10^{12} carbon atoms are C-14 atoms; the rest are C-12. Because of the extra two neutrons, C-14 is slightly radioactive, breaking down into Nitrogen-14 (7 protons and 7 neutrons, one of the C-14 neutrons decays into a proton). C-14 has a half-life of ~5700 years (in 5700 years, half of the original C-14 will have changed into N-14). C-14 combines with oxygen in the atmosphere to form carbon dioxide, just like C-12 does. Carbon dioxide gets fixed into the biomass during photosynthesis so that plants and animals have roughly the same amounts of C-14 in their bodies as the atmosphere has. When the organism dies, there is no more uptake of C-14 which continues to decay. By measuring the amount of C-14 remaining in an organic sample, the age of the sample can be dated. Because of the tiny amounts of C-14 and its half-life, C-14 dating only works back to about 50K-60K years. At 50K-60K years, there are too few C-14 atoms to detect. The above process depends critically upon two assumptions: (1) the sample is not contaminated by more modern C-14 that has not decayed yet; and (2) the initial amount of C-14 in the atmosphere at the time of an organism's death is known. C-14 is formed in the upper atmosphere when N-14 atoms are struck by cosmic rays. Not unexpectedly, the cosmic ray environment will affect the amount of C-14 formed in the atmosphere. We have discussed the fact that the earth has not always been the third planet in our present solar system. What was the cosmic ray background near Kolob? This fact may obviate any and all radioactive dating methods!

Old-Earth or Young-Earth?

Where does the above information lead us? It leads us to the fact that there are enormous unknowns respecting the HOW, WHEN, and HOW LONG of the Creation. The scriptures give some guidance that substantially affects the science of the Creation.

If we assume that 'days' of the pre-Garden Creation are long periods of time, we can tentatively reconcile the scriptures and modern science on some points. For the moment, let us focus on the Old-Earth paradigm. As an example, let us examine how long science theorizes it took to for the Earth's atmosphere to form. According to scientists, the Earth initially had no atmosphere. As volcanoes formed, the Earth's atmosphere became rather toxic to life as we know it. The early atmosphere contained hydrogen sulfide, methane, lots of carbon dioxide, and little to no oxygen. About 4.0 B-2.5 B years ago (a rather large uncertain time frame), the Earth had cooled down enough to enable liquid water to form. Blue-green bacteria (also called cyanobacteria or blue-green algae) are subsequently formed in the oceans. The newly formed bacteria began to release oxygen into the atmosphere using carbon dioxide, water, sunlight, and photosynthesis (a tremendous complex chemical process). The increased oxygen reacted with methane, removing it from the atmosphere. The sky then turned blue. This happened between 2.6B-400M years ago (another rather large uncertain time frame). Evolutionists believed the increased oxygen levels triggered the Cambrian explosion around 520 B years ago. In summary, modern scientists believe it took roughly 3.5 B years to form an oxygen atmosphere![275] After that, ~400M-500M years ago, things moved along comparatively quickly from single-cell, cyanobacteria to humans and all the other advanced life-forms. Consequently, the Old-Earth paradigm may allow for human evolution, allowing some 60M years for homo sapiens to arise after the first appearance of primates. A young-earth paradigm obviously does not allow enough time for Darwinian evolution to work.

Table 3 depicts the proposed modern scientific timeline of events surrounding the naturalistic (Old-Earth) origin of the earth and life. The table is greatly simplified to highlight only the key events. The basic timeline is in MY (millions of years) format. Consequently, 4.6 B years is shown as 4600 MY. The left table shows events primarily in 200 MY steps with some 100 MY steps

[275] https://forces.si.edu/atmosphere/02_02_00.html

used, as necessary. The right table depicts events in 20 MY intervals with one 10 MY interval at the bottom. The right table is, therefore, a 10X zoomed-in image of the bottom of the left table. The table was built consulting over a half-dozen sources, all of which differed – some wildly. It seems the only agreed-upon dates are the age of the earth, the Great Oxidation Event, and the Cambrian Explosion. For consideration, I put 'Day 1' to 'Day 6' in the table to indicate events that could correspond to the 'days' in the Biblical creation record.

Several things in the table are interesting to point out. It took between 200-800 MY for the first prokaryote (the spontaneous generation of life) to appear after liquid water formed. We previously discussed the issues associated with the origin of life. It then only took 100 MY for photosynthesis to appear after the first atmospheric oxygen arose. Photosynthesis is an extremely complicated process requiring molecular machinery more complicated than replisomes and spliceosomes and more on the level of the human retina. It took about 1400 MY for cyanobacteria to appear after the first life form. Eukaryotes then took another 600 MY to appear.

According to evolutionists, the appearance of life and photosynthesis occurred significantly faster than they would expect from chance-based, natural processes. It also seems that the arrival of cyanobacteria and eukaryotes took longer than one might expect. What is most interesting is that plants, flowers, dinosaurs, birds, and mammals arose within ~250 MY of the Cambrian Explosion. It took 10X that long to go from cyanobacteria to single-celled eukaryotes! Plants and animals contain trillions of specialized cells incorporating all kinds of complicated molecular machinery. Primates first appeared ~60 MYA with apes appearing ~50 MY after that. Homo sapiens, the most complicated organism to ever live, presumably occurred ~9.8 MY after apes.

Million Years Ago (100M intervals)	Proposed Event	'Day'
4600	Solar system & the Earth	DAY 1
4500	Skies clear, lights visible	DAY 2
4400	Liquid water	DAY 3
4300	Prokaryotes (single-celled)	
4200		
4100		
4000		
3900		
3800		
3700		
3600		
3500	Earliest Oxygen	
3400	first microbe photosynthesis	
3300		
3200		
3100		
3000		
2900		
2800		
2700	cyanobacteria	
2600	(blue-green algae)	
2500		
2400	GOE: Great Oxidation Event	
2300	(atmosphere changed from	
2200	weak reducing atmosphere	
2100	to oxidizing atmosphere)	
2000		
1900		
1800		
1700	Eukaryotes (single-celled)	
1600	Eukaryotes (multi-celled)	
1500		
1400	Sexual reproduction	
1300		
1200		
1100		
1000		
900		
800	Green algae	
700	Earliest plants	DAY 4
600	Earliest animals	DAY 5
500	Cambrian Explosion	
400		
300		
200		
100		
0	present day	

Million Years Ago (10M intervals)	Proposed Event	'Day'
550		
540	Cambrian Explosion began	
530		
520	Cambrian Explosion peak	
510	arthropods, molluscs	
500		
490		
480		
470		
460		
450		
440		
430		
420		
410		
400		
390		
380		
370		
360		
350	plants	
340		
330		
320		
310		
300		
290		
280	dinosaurs	
270		
260		
250		
240		
230		
220	birds	
210		
200	mammals	
190		
180	flowers	
170		
160		
150		
140		
130		
120		
110		
100		
90		
80		
70		
60	primates	
50		
40		
30		
20		
10	apes	
0	present day	DAY6

LEGEND:
MY = Million Years
1000 MY = Billion Years
0.2 MY = 200 Thousand Years

right table expanded by 10X vs. left table

200K years/modern humans →

TABLE 3. A TIMETABLE OF THE POSSIBLE EVENTS SURROUNDING THE CREATION OF THE EARTH AND OF LIFE ACCORDING TO MODERN SCIENCE.

In addition to the Young-Earth and Old-Earth paradigms, there is at least a third option. I will call this third paradigm the Old-Enough-Earth-Preparation paradigm. In this model, God used natural processes to prepare the Earth for modern plants, animals, and humans. Subsequently, all life would have to be purged from the Earth before the Garden of Eden was prepared when modern plant seeds from other worlds were planted and modern animals from other worlds procreated to fill the Garden with life. These plants and animals were immortal and remained so until the Fall of Adam and Eve. In the Fall, Adam and Eve became the first 'flesh' on the prepared Earth (they became mortal).

In the end, how long the Creation took is of little consequence. Originally, I favored the Old-Enough-Earth-Preparation paradigm. With additional study and time, I have come to favor a Young-Earth paradigm or at least a Younger Earth – primarily as a result of taking the scriptures more literally and knowing that, due to cataclysms, things can happen very rapidly on the earth. The important thing to remember is God's foreknowledge in providing the Plan of Exaltation and an Earth on which to carry out that plan. A means had to be provided for Adam and Eve and the entirety of the immortal creation to become mortal, hence, the Tree of the Knowledge of Good and Evil. To compensate for physical and spiritual death concomitant with mortality and the Fall of Man, the Plan required a Savior. Our Elder Brother in the spirit world, Jesus Christ, stepped forward and volunteered to take upon Him that role at Father-in-Heaven's request. Nothing was left to chance. **Clearly, the origin of man was planned and carried out by the Gods. Man is a direct descendant of God. This fact does not denigrate God but immensely elevates man**.

Does it make sense that a blind, purposeless, chance-based process without a time limit was used to create the bodies of all the other life-forms besides man as well? What about the origin of life itself? Or the origin of natural laws themselves? Science is fond of claiming that they use Occam's Razor. **Occam's Razor states that if there are two explanations for some phenomenon, the simpler explanation is generally the correct one.** The competing explanations are Darwinian evolution and Intelligent Design. Occam's Razor overwhelmingly favors ID!

Let us conclude this chapter with Rodney Turner's declaration. ***"The earth did not create man; [an exalted] Man created the earth. Earth is not where man first set his foot, but where he last set his foot. His origin is not down in the dust, but up in the stars. He is not the offspring of primates, but of the Gods. His ever-receding origin is not to be found in East Africa, but on Kolob – and beyond."***[276]

---

[276] Turner, The Footstool of God, p. 77

11. CONCLUSION

This treatise began with two contrasting worldviews – that of the Prophet Joseph Smith versus that of the scientist Charles Darwin. Darwin's worldview insists that all of nature is the result of a purely natural process which "operates by chance, is random, unguided, blind, and purposeless."[277] The Prophet's worldview, on the other hand, believes that the "creation of or the intelligent design of life is deliberately designed, planned, revealed, and purposeful."[278] The thesis of this paper is that Joseph Smith's worldview is the correct view. The Lord revealed His plan and the purpose for mankind to The Prophet.

"That by reason of transgression cometh the fall, which fall bringeth death, and inasmuch as ye were ***born into the world by water****, and* ***blood****, and the* ***spirit***, which I have made**,** and so became of **dust a living soul**, even so ye must be ***born again into the kingdom of heaven***, of ***water***, and of the ***Spirit***, and be cleansed by ***blood***, even the blood of mine Only Begotten; that ye might be sanctified from all sin, and enjoy the words of eternal life in this world, and ***eternal life in the world to come, even immortal glory***."[279]

Darwin is no Joseph Smith. In the spring of 1820, in answer to prayer, God the Father and His Son, Jesus Christ, appeared to the fourteen-year-old Joseph Smith in the Sacred Grove. Several years later, in answer to prayer, the Angel Moroni appeared to Joseph Smith three times on the evening of September 21, 1823, and once the next day. Four years later, Moroni gave Joseph the plates from which the Book of Mormon was to be translated. During the intervening four years, the boy Prophet received more than 88 angelic visitations from more than 50 distinct angels. Among these

---

[277] Introduction

[278] Introduction

[279] Moses 6:59

visits were at least three from Father Adam and one for Mother Eve.[280] The principles that the Prophet Joseph Smith, Jr., taught came directly from the Almighty God.

Regrettably, many evolutionary evangelists meet every criterion of an Anti-Christ, denying the existence of God, the Atonement of Christ, the doctrines of exaltation, and revelation. By denying God and revelation and anything they cannot 'see' with their senses, they ask for signs. In replying to Korihor's demand that Alma, the prophet, give him a sign, Alma boldly stated, "Thou hast had signs enough; will ye tempt your God? Will ye say, show unto me a sign, when **ye have the testimony of all these thy brethren, and also all the holy prophets**? **The scriptures are laid before thee**, yea, and ***all things denote there is a God; yea, even the earth, and all things that are upon the face of it, yea, and its motion, yea, and also all the planets which move in their regular form do witness that there is a Supreme Creator***." [281] The world has adopted Darwinian evolution whole cloth and have extended its premises to numerous completely unrelated fields such as economics, psychology, the human mind, and human society – with disastrous results! "**The influence of evolutionary theory on fields far removed from biology is one of the most spectacular examples in history of how a highly speculative idea for which there is no really hard scientific evidence can come to fashion thinking of a whole society and dominate the outlook of an age**."[282]

---

[280] Opening the Heavens: Accounts of Divine Manifestations, 1820-1844, BYU Press, 2005, *Parting the Veil: Joseph Smith's Seventy-Six Documented Visionary Experiences*, Alexander L. Baugh, pp. 265-326; Pres. John Taylor - JD 17:374; 18:325-6; 20:174-5; 21:65, 94, 161-163; 23:48-49; Pres. Wilford Woodruff - JD 16:265-7; Elder Orson Pratt - JD 13:67; 15:185; Pres. George Q. Cannon - JD 13:47; 23:362)

[281] Alma 30:44

[282] Denton, Evolution: A Theory in Crisis, 1985, p. 384

It seems that many scientists are "Ever learning and never able to come to the knowledge of the truth."[283] Thomas S. Kuhn has written a seminal book on The Structure of Scientific Revolutions,[284] **Kuhn describes scientific "research as a strenuous and devoted attempt to force nature into conceptual boxes supplied by a professional education."**[285] In today's scientific world, one must bow to Darwinian evolution or be rejected by their scientific peers and be rejected for research funds. Kuhn coined the term 'paradigms' which means a set of scientific concepts or hypotheses. We have all been told that science is ever searching for the truth, accepting each new verifiable idea with enthusiasm. Unfortunately, this is seldom the case. When describing how new theories come to be accepted, Kuhn explains, "Conversions [of scientists to new paradigms} will occur a few at a time until, after the last holdouts have died, the whole profession will again be practicing under a single, but now a different, paradigm."[286] In other words, **new theories come into play not because of overwhelming evidence, but because the proponents of the old theories – die**.

Speaking at the 1974 annual conference of the American Association for the Advancement of Science (AAAS), Carl Sagan, a noted scientist, and an avowed evolutionist explained how science purportedly works. **Carl Sagan said, "(1) the most fundamental axioms and conclusions may be challenged; (2) the prevailing hypothesis must survive a confrontation with observation; (3) appeals to authority ... are impermissible; and (4) experiments must be reproducible."** Unfortunately, when it comes to evolution,

---

[283] 2 Tim. 3:7

[284] Kuhn, The Structure of Scientific Revolutions, 50th Anniversary Edition, 4th Ed., 2012

[285] Kuhn, The Structure of Scientific Revolutions, p. 5

[286] Kuhn, The Structure of Scientific Revolutions, p. 152

the archetypical scientific method is modified. Marvin Lubenow observes that **no one in the scientific establishment is permitted to challenge evolution, evolution is non-observable, evolutionists often appeal to authority, and there are no repeatable experiments that can confirm evolution**.

Modern science is laboring under the false hopes of Darwinian evolution. While it can explain some minor adaptations, it cannot explain macroevolution or the origin of life. Yet, despite the lack of evidence, most scientists cling to Darwinian evolution as their only hope – to the point of irrationality. At the end of his book, Darwin Devolves, Behe posits that modern science has started to spurn reason. "Without the fundamental underpinnings of reason – science itself eventually goes off the rails, plunging deeper and deeper into irrationality."[287] Behe relates the history of an oddball idea. In 2002, Nick Bostrom, an Oxford University professor of philosophy, wrote a book entitled Anthropic Bias: Observation Selection Effects in Science and Philosophy.[288] In the book, Bostrom discussed why our universe is so finely tuned for life. There are two hypotheses, the 'design hypothesis' and the 'ensemble hypothesis.' The ensemble hypothesis states that there are many multiverses. Since our universe contains life, the laws in this universe **must be** finely tuned for life. Bostrum basically ignored the design hypothesis. The next year, he published an article, *Are We Living in a Computer Simulation?* [289] In that article, Bostrum argues that mankind is living in a computer simulation. The idea took on a life of its own. Thirteen years later, The American Museum of Natural History hosted the 2016 Isaac Asimov Memorial Debate: Is the Universe a Simulation? The chair of the event was a science-popularizing astrophysicist, Neil deGrasse

---

[287] Behe, Darwin Devolves, p. 270

[288] Nick Bostrum, Anthropic Bias: Observation Selection Effects in Science and Philosophy, Routledge, 2010

[289] Bostrum, Philosophical Quarterly, 53, 2003, pp. 243-255

Tyson. Tyson commented during the event, "I think the likelihood [of mankind living in a computer simulation] may be very high." The absurdity of the idea would seem to have been enough to stop the idea in its tracks. Instead, some of America's premier institutions and scientists came to give credence to and nurture the idea as being a serious idea. **Irrationality is beginning to reign – all because science cannot accept that there is an Intelligence behind all of nature**. If we do live in a computer simulation, who is running the simulation? Who designed the computer? This sounds a lot like an Intelligent Design argument.

In Gelernter's essay, he comments that "***Darwin is no Newton***. Newton's physics survived Einstein and will always survive, because it explains the cases that dominate all of space-time except ... at the very smallest and largest scales. **It's just these most important cases, the ones we see all around us that Darwin cannot explain**."[290] [emphasis in the original] The most important cases in evolution involve macroevolution. At the end of his essay, Gelernter wonders if biology can "get over Darwin, and move on?" This "is one of the most important questions facing science in the 21st century."

The Prophet spoke the Lord's doctrine. ***The Lord's doctrine specifies that MAN IS A SPECIAL CREATION and that THERE IS A PURPOSE FOR OUR EXISTENCE.***

When Charles Darwin was buried next to Isaac Newton in Westminster Abbey, Newton may have started to rotate on his own axis. While Darwin abandoned his study for the ministry to create a theory that is the very foundation of modern materialism[291], "Newton spent a greater part of his life studying theology than science. ... It was Newton's belief that an intelligent Creator had created a rational world, with a rational order, that gave him the

---

[290] Gelernter, *Giving Up Darwin*, Claremont Review of Books, 2019, p. 109

[291] Behe, Darwin Devolves, p. 273

confidence to experiment with mathematical formulas to describe and predict how that world functions. ... **Newton was also convinced that science gave evidence of God's existence, because he saw design in the cosmos and in our own bodies**. ... In Newton's words, the purpose of science is to 'deduce causes from effects, until we come to the very first cause, which certainly is not mechanical' — in other words, a Creative Mind,"[292] i.e., The Great and Last Clockmaker.

In conclusion, the Lord gave Moses a vision of the Earth and its inhabitants.

> "Moses beheld the world and the ends thereof, and all the children of men which are, and which were **created**; of the same he greatly marveled and wondered."[293]

After the Lord left Moses, he was unable to move for the space of many hours. After experiencing the presence of the Lord, Moses exclaimed,

> "Now, for this cause I know that man is nothing, which thing I never had supposed."[294]

Moses then had a second vision of the Lord.

> "And the Lord God said unto Moses: **For mine own purpose have I made these things**.
>
> "And by the word of my power, have I created them, which is mine Only Begotten Son, who is full of grace and truth.
>
> "And **worlds without number have I created**; and I also created them for mine own purpose; and **by the Son I created them, which is mine Only Begotten**.

---

[292] Tina Simmons, *Isaac Newton in Quarantine*, American Thinker, May 2, 2020

[293] Moses 1:8

[294] Moses 1:10

> "And the ***first man of all men have I called Adam***, which is many.
>
> "But **only an account of this earth, and the inhabitants thereof, give I unto you**. For behold, **there are many worlds** that have passed away by the word of my power. And there are many that now stand, and innumerable are they unto man; but all things are numbered unto me, for they are mine and I know them.
>
> ***"For behold, this is my work and my glory—to bring to pass the immortality and eternal life of man***.
>
> "And now, Moses, my son, I will speak unto thee concerning this earth upon which thou standest; and **thou shalt write the things which I shall speak**.
>
> **"And in a day when the children of men shall esteem my words as naught [the Last Days] and take many of them from the book which thou shalt write [the Bible], behold, I will raise up another like unto thee [the Prophet Joseph Smith]; and they shall be had again among the children of men—among as many as shall believe**."[295]

After restoring Moses' words, the Prophet Joseph Smith had a vision. In recording his vision, the Prophet wrote,

> "And now, after the many testimonies which have been given of him, this is the testimony, last of all, which we give of him: ***That he lives!***
>
> For we saw him, even on the right hand of God; and we heard the voice bearing record that he is the Only Begotten of the Father—

---

[295] Moses 1:31-35, 39-41

***That by him, and through him, and of him, the worlds are and were created, and the inhabitants thereof are begotten sons and daughters unto God*.”[296]**

[296] D&C 76:22-24

APPENDIX 1: The Origin of Man

The year 1909 was the centennial year of Charles Darwin's birth and the 50th anniversary of his publication On the Origin the Species. The preface to the 1909 Doctrinal Exposition by the First Presidency, *The Origin of Man*, reads as follows. *"In the early 1900s, questions concerning the Creation of the earth and the theories of evolution became the subject of much public discussion. In the midst of these controversies, the First Presidency issued the following in 1909, which expresses the Church's doctrinal position on these matters."* This exposition was published in The Improvement Era, Nov. 1909. The exposition was signed by Presidents Joseph F. Smith, John R. Winder, and Anthon H. Lund.

FIGURE 17. FIRST PRESIDENCY OF THE CHURCH OF JESUS CHRIST OF LATTER-DAY SAINTS, NOVEMBER 1909

Soon thereafter, in 1925, the United States of America was enduring the Scopes 'Monkey Trial.' The case centered on John T. Scopes, a high school teacher, who presumably taught human evolution in his classroom contrary to a recent Tennessee state law. The First Presidency then issued a second Doctrinal Exposition entitled, *Mormon View of Evolution*. This exposition was a word-for-word extraction from the *Origin of Man* exposition. The second exposition was signed by Presidents Heber J. Grant, Anthony W. Ivins, and Charles W. Nibley. It was published in The Improvement Era, Sep. 1925.

FIGURE 18. FIRST PRESIDENCY OF THE CHURCH OF JESUS CHRIST OF LATTER-DAY SAINTS, SEPTEMBER 1925

The full text of the *Origin of Man* is found below. The word-for-word extracts comprising the *Mormon View of Evolution* are underlined. Capitalization, punctuation, paragraphing, and spelling have been standardized. [Emphasis has been added by the author.]

< BEGINNING OF EXPOSITIONS >

"God created man in his own image." (Gen. 1:27)

Inquiries arise from time to time respecting the attitude of The Church of Jesus Christ of Latter-day Saints upon questions which, though not vital from a doctrinal standpoint, are closely connected with the fundamental principles of salvation. The latest inquiry of this kind that has reached us is in relation to the origin of man. It is believed that a statement of the position held by the Church upon this subject will be timely and productive of good.

In presenting the statement that follows we are not conscious of putting forth anything essentially new; neither is it our desire so to do. Truth is what we wish to present, and truth—eternal truth—is fundamentally old. A restatement of the original attitude of the Church relative to this matter is all that will be attempted here. To tell the truth as God has revealed it, and commend it to the acceptance of those who need to conform their opinions thereto, is the sole purpose of this presentation.

"God created man in his own image, in the image of God created he him; male and female created he them." In these plain and pointed words the inspired author of the book of Genesis made known to the world the truth concerning the origin of the human family. Moses, the prophet-historian—"learned," as we are told, "in all the wisdom of the Egyptians"—when making this important announcement was not voicing a mere opinion, a theory derived from his researches into the occult lore of that ancient people. **He was speaking as the mouthpiece of God, and his solemn declaration was for all time and for all people**. No subsequent revelator of the truth has contradicted the great leader and lawgiver of Israel. All who have since spoken by divine authority upon this theme have confirmed his simple and sublime proclamation. Nor could it be otherwise. Truth has but one source, and all revelations from heaven are harmonious with each other. **The omnipotent Creator, the maker of heaven and earth, had shown unto Moses everything pertaining to this planet, including the facts relating to man's origin**, and the authoritative pronouncement of that mighty prophet and seer to the house of Israel, and through Israel to the whole world, is couched in the simple clause: "God created man in his own image" (Gen. 1:27; see Moses 1:27–41).

The creation was twofold—first spiritual, secondly temporal. This truth, also, Moses plainly taught—much more plainly than it has come down to us in the imperfect translations of the Bible that are now in use. Therein the fact of a spiritual creation, antedating the temporal creation, is strongly implied, but the proof of it is not so clear and conclusive as in other records held by the Latter-day Saints to be of equal authority with the Jewish scriptures. The partial obscurity of the latter upon the point in question is owing, no doubt, to the loss of those "plain and precious" parts of sacred writ, which, as the Book of Mormon informs us, have been taken away from the Bible during its passage down the centuries (see 1 Ne. 13:24–29). Some of these missing parts the Prophet Joseph

Smith undertook to restore when he revised those scriptures by the spirit of revelation, the result being that more complete account of the Creation which is found in the book of Moses, previously cited. Note the following passages:

"And now, behold, I say unto you, that these are the generations of the heaven and of the earth, when they were created, in the day that I, the Lord God, made the heaven and the earth,

"And every plant of the field before it was in the earth, and every herb of the field before it grew. For I, the Lord God, created all things of which I have spoken, spiritually, before they were naturally upon the face of the earth. For I, the Lord God, had not caused it to rain upon the face of the earth. And I, the Lord God, had created all the children of men; and not yet a man to till the ground; for in heaven created I them, and there was not yet flesh upon the earth, neither in the water, neither in the air;

"But I, the Lord God, spake, and there went up a mist from the earth, and watered the whole face of the ground.

"And I, the Lord God, formed man from the dust of the ground, and breathed into his nostrils the breath of life; and man became a living soul, the **first flesh upon the earth**, the **first man also**; nevertheless, all things were before created; but spiritually were they created and made according to my word" (Moses 3:4–7; see also Moses 1 and Moses 2, and compare with Gen. 1 and Gen. 2).

These two points being established, namely, the creation of man in the image of God, and the twofold character of the Creation, let us now inquire: **What was the form of man, in the spirit and in the body, as originally created?** In a general way the answer is given in the words chosen as the text of this treatise. "God created man in his own image." It is more explicitly rendered in the Book of Mormon thus: ***"All men were created in the beginning after mine own image"*** (Ether 3:15). ... If, therefore, we can ascertain the form of the "Father of spirits," "The God of the spirits of all flesh," we shall be able to discover the form of the original man.

Jesus Christ, the Son of God, is "the express image" of His Father's person (Heb. 1:3). He walked the earth as a human being, as a perfect man, and said, in answer to a question put to Him: "He that hath seen me hath seen the Father" (John 14:9). This alone ought to solve the problem to the satisfaction of every thoughtful, reverent mind. The conclusion is irresistible, that **if the Son of God be the express image (that is, likeness) of His Father's person, then His Father is in the form of a man; for that was the form of the Son of God, not only during His mortal life, but before His mortal birth, and after His Resurrection**. It was in **this form that the Father and the Son, as two personages, appeared to Joseph Smith**, when, as a boy of 14 years, he received his first vision. Then if God made man—the first man—in His own image and likeness, He must have made him like unto Christ, and consequently like unto men of Christ's time and of the present day. That man was made in the image of Christ is positively stated in the book of Moses: "And I, God, said unto mine Only Begotten, which was with me from the beginning, Let us make man in our image, after our likeness; and it was so. ...

"And I, God, created man in mine own image, in the image of mine Only Begotten created I him; male and female created I them" (Moses 2:26–27).

The Father of Jesus is our Father also. Jesus Himself taught this truth when He instructed His disciples how to pray: "Our Father which art in heaven," etc. **Jesus, however, is the firstborn among all the sons of God—the first begotten in the spirit, and the only begotten in the flesh. He is our elder brother, and we, like Him, are in the image of God**. ***All men and women are in the similitude of the universal Father and Mother and are literally the sons and daughters of Deity.***

"God created man in His own image." This is **just as true of the spirit as it is of the body**, which is only the clothing of the spirit, its complement—the two together constituting the soul. **The spirit of man is in the form of man**, and the spirits of all creatures are in the likeness of their bodies. This was plainly taught by the Prophet Joseph Smith (see D&C 77:2; "that which is spiritual being in the likeness of that which is temporal; and that which is temporal in the likeness of that which is spiritual; the spirit of man in the likeness of his person").

Here is further evidence of the fact. More than 700 years before Moses was shown the things pertaining to this earth, another great prophet, known to us as the brother of Jared, was similarly favored by the Lord. He was even permitted to behold the spirit-body of the foreordained Savior, prior to His incarnation; and so, like the body of a man was gazing upon a being of flesh and blood. He first saw the finger and then the entire body of the Lord—all in the spirit. The Book of Mormon says of this wonderful manifestation:

"And it came to pass that when the brother of Jared had said these words, behold, the Lord stretched forth his hand and touched the stones one by one with his finger. And the veil was taken from off the eyes of the **brother of Jared**, and he **saw the finger of the Lord; and it was as the finger of a man**, like unto flesh and blood; and the brother of Jared fell down before the Lord, for he was struck with fear.

"And the Lord saw that the brother of Jared had fallen to the earth; and the Lord said to him: Arise, why hast thou fallen?

"And he saith unto the Lord: I saw the finger of the Lord, and I feared lest he should smite me; for I knew not that the Lord had flesh and blood.

"And the Lord said unto him: Because of thy faith thou hast seen that I shall take upon me flesh and blood; and never has man come before me with such exceeding faith as thou hast; for were it not so ye could not have seen my finger. Sawest thou more than this?

"And he answered: Nay; Lord, show thyself unto me.

"And the Lord said unto him: Believest thou the words which I shall speak?

"And he answered, Yea, Lord, I know that thou speakest the truth, for thou art a God of truth, and canst not lie.

"And when he had said these words, behold, the Lord showed himself unto him, and said: Because thou knowest these things ye are redeemed from the fall; therefore ye are brought back into my presence; therefore I show myself unto you.

"Behold, I am he who was prepared from the foundation of the world to redeem my people. Behold, I am Jesus Christ. I am the Father and the Son. In me shall all mankind have life, and that eternally, even they who shall believe on my name; and they shall become my sons and my daughters.

"And never have I showed myself unto man whom I have created, for never has man believed in me as thou hast. ***Seest thou that ye are created after mine own image? Yea, even all men were created in the beginning after mine own image***.

"Behold, this body, which ye now behold, is the body of my spirit; and man have I created after the body of my spirit; and even as I appear unto thee to be in the spirit will I appear unto my people in the flesh" (Ether 3:6–16).

What more is needed to convince us that ***man, both in spirit and in body, is the image and likeness of God and that God Himself is in the form of a man***?

When the divine Being whose spirit-body the brother of Jared beheld took upon Him flesh and blood, He appeared as a man, having "body, parts and passions," like other men, though vastly superior to all others, because He was God, even the Son of God, the Word made flesh: in Him "dwelt the fulness of the Godhead bodily." And why should He not appear as a man? That was the form of His spirit, and it must needs have an appropriate covering, a suitable tabernacle. He came into the world as He had promised to come (see 3 Ne. 1:13), taking an infant tabernacle and developing it gradually to the fulness of His spirit stature. He came as man had been coming for ages and as man has continued to come ever since. Jesus, however, as shown, was the Only Begotten of God in the flesh.

Adam, our first progenitor, "**the first man**," was, **like Christ, a preexistent spirit**, and **like Christ he took upon him an appropriate body, the body of a man**, and so became a "living soul." The **doctrine of the preexistence—revealed so plainly, particularly in latter days—pours a wonderful flood of light upon the otherwise mysterious problem of man's origin**. It shows that man, as a spirit, was begotten and born of heavenly parents and reared to maturity in the eternal mansions of the Father, prior to coming upon the earth in a temporal body to undergo an experience in mortality. It teaches that all men existed in the spirit before any man existed in the flesh and that **all who have inhabited the earth since Adam have taken bodies and become souls in like manner**.

It is **held by some** that Adam was not the first man upon this earth **and that the original human being was a development from lower orders of the animal creation**. ***These, however, are the theories of men. The word of the Lord declared that Adam was "the first man of all men" (Moses 1:34), and we are therefore in duty bound to regard him as the primal parent of our race***. It was shown to the brother of Jared that ***all men were created in the beginning after the image of God; whether we take this to mean the spirit or the body, or both***, it commits us to the same conclusion: ***Man began life as a human being, in the likeness of our Heavenly Father***.

True it is that the body of man enters upon its career as a tiny germ embryo, which becomes an infant, quickened at a certain stage by the spirit whose tabernacle it is, and the child, after being born, develops into a man. There is nothing in this, however, to indicate that the original man, the first of our race, began life as anything less than a man, or less than the human germ or embryo that becomes a man.

Man, by searching, cannot find out God. Never, unaided, will he discover the truth about the beginning of human life. The Lord must reveal Himself or remain unrevealed; and the same is true of the facts relating to the origin of Adam's race—God alone can reveal them. Some of these facts, however, are already known, and what has been made known it is our duty to receive and retain.

The Church of Jesus Christ of Latter-day Saints, basing its belief on divine revelation, ancient and modern, proclaims man to be the direct and lineal offspring of Deity. God Himself is an exalted man, perfected, enthroned, and supreme. By His almighty power He organized the earth and all that it contains, from spirit and element, which exist coeternally with Himself. He formed every plant that grows and every animal that breathes, each after its own kind, spiritually and temporally—"that which is spiritual being in the likeness of that which is temporal, and that which is temporal in the likeness of that which is spiritual." He made the tadpole and the ape, the lion, and the elephant, but He did not make them in His own image, nor endow them with godlike reason and intelligence. Nevertheless, the whole animal creation will be perfected and perpetuated in the Hereafter, each class in its "distinct order or sphere," and will enjoy "eternal felicity." That fact has been made plain in this dispensation (see D&C 77:3).

Man is the child of God, formed in the divine image and endowed with divine attributes, and even as the infant son of an earthly father and mother is capable in due time of becoming a man, so the undeveloped offspring of celestial parentage is capable, by experience through ages and aeons, of evolving into a God.

< END OF EXPOSITIONS >

It has been suggested that because the above statement in blue, referring to the 'theories of men,' was not included in the second Doctrinal Exposition by the First Presidency, the anti-science statement was no longer valid or, at the least, deemphasized. If by exclusion, statements are no longer valid or have been 'softened,' what about the above statement in green? '***God Himself is an exalted man, perfected, enthroned, and supreme***.' This is a restatement of a doctrine taught by Lorenzo Snow and confirmed by the Prophet, "As man now is, God once was; As God now is, man may be."[297] This doctrine was later affirmed in the King Follett Discourse in 1844. The discourse was given by the Prophet Joseph Smith at King Follett's funeral in Nauvoo, Illinois.[298] It appears that the omission was more in the interest of brevity and possibly de-escalating passions than for any doctrinal reason.

The early 1900s were filled with discussion about the place of science in society. In 1910, the year after the *Origin of Man* exposition was given, there was a controversy at Brigham Young University (BYU) referred to as the Modernism Controversy.[299] Four professors at BYU were teaching evolution and higher criticism of the Bible. As a result, the First Presidency referred to the Church's position on science during their Christmas message. "It stated that the church is not hostile to science and … that **proven science is accepted with joy, but theories, speculation, or**

---

[297] Snow, LeRoi C., Improvement Era, June 1919, p. 656

[298] Smith, Joseph Fielding, The Teachings of the Prophet Joseph Smith, Deseret Book (1938), pp. 345-6; see also Stan Larson, *The King Follett Discourse: A Newly Amalgamated Text*, BYU Studies 18, No. 2 (1978)

[299] Wikipedia: 1911 Brigham Young University modernism controversy; https://en.wikipedia.org/wiki/1911_Brigham_Young_University_modernism_controversy

anything contrary to revelation or common sense are not accepted."[300]

The emphasized sections of the Doctrinal Expositions are crystal clear on the origin of man: "***Man BEGAN life as a human being, in the likeness of our Heavenly Father,***" both in the spirit and in the body.

---

[300] Wikipedia: Mormon views on evolution; https://en.wikipedia.org/wiki/Mormon_views_on_evolution#:~:text=The%20LDS%20Church%20has%20no,evolution%20is%20opposed%20to%20scriptural

APPENDIX 2: BYU's Monte L. Bean Museum Evolution Display

Understanding the Principles of Evolution

In the BYU Magazine, Summer 2019, there is an article entitled *Evolution on Display at BYU.*[301] In the magazine, BYU College of Life Sciences, Summer/Fall 2019, there is a similar article entitled the *Monte L. Bean Museum: Understanding the Principles of Evolution*.[302] The latter article states that the new exhibit in the museum is part of a growing trend "towards reconciling religion and evolution." The museum director said, "The heart of our mission here at the Bean Museum is to **celebrate the role of Jesus Christ as the creator of life on Earth**, and ***to use science to understand His works better***." Jamie L. Jensen and Seth M. Bybee, evolutionary biologists at BYU, helped put the evolution exhibit together. After suggesting that "humans share a common ancestor with apes," they embarrassingly state that their "main motivation is [to] help save ... testimonies." Jensen explains, "We're out to help find ways that they [Church members] can keep their faith and still be knowledgeable." Ethan Tolman, a student researcher, says the exhibit helps museum patrons approach evolution with "an open mind." The implication is that if someone does NOT accept evolution, they do **not** have an OPEN MIND and they are **not** KNOWLEDGEABLE! **I would suggest we use the scriptures and modern prophets instead of science to understand Christ's works better.**

The Monte L. Bean Museum website contains the following text. "**Nothing has been revealed concerning evolution**. Though **the details of what happened on earth before Adam and Eve, including how their bodies were created, have not been revealed, *our teachings regarding man's origin are clear and come from revelation***. The exhibit, *Understanding the Principles of Evolution,*

---

[301] Kristen L. Evans, *Evolution on Display at BYU*, BYU Magazine, Summer 2019, p. 12

[302] Amelia Mecham, *The Monte L. Bean Museum: Understanding the Principles of Evolution*, BYU College of Life Sciences, Summer/Fall 2019, p. 14

is an exploration of how [some] scientists interpret the theory of evolution and does not establish the doctrine of The Church of Jesus Christ of Latter-day Saints. **Visitors are encouraged to consider how natural selection plays a part in nature to help animals, plants, and other life-forms ADAPT**."[303] [emphasis added] This entire paragraph is deceitful. The object of the display is clearly designed to influence the interpretation of Church doctrine to conform to some scientists' views on evolution, particularly the postulated evolution of mankind.

There are numerous problems with this exhibit, the text associated with it, and their attempt to reconcile science and religion, specifically, the doctrine of The Church of Jesus Christ of Latter-day Saints. First, the museum states that its goal is to explain "how natural selection [helps] life-forms [to] adapt." Adaptation is a microevolutionary process, not a macroevolutionary process. Second, roughly half of the museum's display is on adaptation, the other half is squarely about the macroevolution paradigm. Third, the website explains that "**nothing** has been revealed concerning evolution." How can they say that when ***two of the Church's three Doctrinal Expositions of the First Presidency deal with the evolution of man***! [304] Fourth, the museum assumes that reconciliation between religion and science on evolution is both possible and good. It appears to me that reconciliation is neither possible nor good since macroevolution is not true based upon the information provided in this treatise. Fifth, BYU is known for discovering and naming giant sauropods and the Utahraptor – there is not one word mentioning those discoveries. Instead, the

---

[303] https://mlbean.byu.edu/exhibits/understanding-the-principles-of-evolution

[304] Encyclopedia of Mormonism, Appendix 4, *The Origin of Man* (1908) and *The Mormon View of Evolution* (1925); the third Doctrinal Exposition of the First Presidency is on *The Father and the Son*, President Joseph F. Smith (1916)

center display is about the man descending from apes, a subject on which BYU has zero experience and is the most controversial of all evolutionary topics. Sixth, I can see no way that organic evolution can be made to reconcile with Church doctrine concerning mankind, and by extension, the rest of His creations.

The contents of the First Presidency Doctrinal Exposition titled The Origin of Man comprises four points.

(1) God created humans.

(2) God created Adam, "the origin of the human family" and "the first man."

(3) The creation was sequential: first spiritual, later physical.

(4) Each human body looks like and displays the characteristics of the individual pre-mortal spirit that inhabits it.

Given the above, it seems disingenuous and deceitful that the BYU's Monte L. Bean Museum's display about *Understanding the Principles of Evolution* does NOT emphasize simple adaptation, but instead, display a speculative arrangement of hominid bones that are designed to give the impression that man descended from apes. See Figure 18. It seems that some of the Church's intellectuals give the doctrines of the Church a tithe of their faith. I am disappointed and heart-broken that Brigham Young University would sponsor such a display. President Brigham Young – along with Isaac Newton – is likely rotating on his axis in his grave given the hubris of this display at his namesake's university. Consider the numerous quotes by Brigham Young in the foregoing treatise – is it conceivable that he would approve of this display?

Let us reason about evolution versus the doctrine of the Creation. Doctrinally, the argument is simple. Our spirit bodies look like God the Father and God the Mother because we were all born of them as spirit children. They provided spirit bodies for our intelligences. With three exceptions, all humans were born of mortal parents. We know that Jesus Christ's earthly body was born of Mary, a mortal woman, and His Father, Elohim, an exalted and glorified man. From his mother, Christ had the power to die. From his Father, he had the power to live forever. What about Adam and Eve? As mentioned in Chapter 10, we have reason to believe that Adam and Eve were born of Heavenly Father and Heavenly Mother in the Garden of Eden. Our Heavenly Father and Mother can have physical children if they so choose. However, those children would be immortal – not glorified, but immortal. Consequently, they would need a means whereby they could become mortal. Hence, the Tree of the Knowledge of Good and Evil in the midst of the Garden. Let us discuss each of these ideas in a little more depth.

The Family Proclamation that states that "all human beings—male and female—are created in the image of God. Each is a beloved spirit son or daughter of heavenly parents." Similarly, the Doctrinal Exposition on the *Origin of Man*, emphatically states, "man, as a spirit, was begotten and born of heavenly parents." While still in the spirit world as a spirit being, Christ appeared to the Brother of Jared. He explained, "This body, which ye now behold, is the body of my spirit; and **man have I created after the body of my spirit**; and **even as I appear unto thee to be in the spirit will I appear unto my people in the flesh**."[305] God the Father has a body of human form. Man is made in His image. "I, God, said unto mine Only Begotten, which was with me from the beginning: **Let us make man in our image**, after our likeness; and it was so."[306] Our spirit creation is a type of our physical creation. "For by the power of my Spirit **created** I them; yea, all things **BOTH spiritual and temporal** — First spiritual, secondly temporal, which is the beginning of my work; and again, first temporal, and secondly spiritual, which is the last of my work."[307] In a revelation regarding the Book of Revelation, the Lord said, "**that which is spiritual being**

---

[305] Ether 3:4-16

[306] Moses 2:26

[307] D&C 29:31-32

in the likeness of that which is temporal; and that which is temporal in the likeness of that which is spiritual; the spirit of man in the likeness of his person."[308] Similarly, "**all things have their likeness**, and all things are created and made to bear record of me, **both things which are temporal, and things which are spiritual**."[309] Doctrinally, if birth was how we obtained our spirit bodies, then birth is how Adam and Eve and the rest of us obtained our physical bodies!

How does evolution fit into the Plan of Exaltation? According to those advocating the evolution of the human family, Adam and Eve had descended from lower animals until they were advanced enough to look like God and have the mind and spirit to desire to become like Him. There are several fundamental problems with this hypothesis. First, Adam and Eve were the first of all flesh, i.e., there was no death until the Fall. They had to partake of the fruit to become mortal. Joseph Fielding Smith wrote, "When Adam was in the Garden of Eden, he was not subject to death."[310] Second, evolution is a random, trial and error process. As such, there is no time limit and no end goal. Reading the scriptures or observing the temple endowment, one is awed by the perfect planning that the Gods went through to 'bring to pass the immortality and eternal life of man.' Does it make sense that God would sit around waiting for evolution to finish? Finish what? Making bodies that just happened to look like God's exalted body and Christ's, Adam's, and Eve's spirit bodies! Remember, evolution can have no goal, it cannot know that it is supposed to produce bodies that look like

---

[308] D&C 77:2

[309] Moses 6:63

[310] Smith, Joseph Fielding, Doctrines of Salvation, 1:76-77; quoted in Doctrines of the Gospel Student Manual, *Chapter 8: The Fall*, published by The Church of Jesus Christ of Latter-Day Saints; https://www.churchofjesuschrist.org/study/manual/doctrines-of-the-gospel-student-manual/8-the-fall?lang=eng

our spirits. Third, ***even if evolution managed by some extraordinary chance make the right kind of bodies for Adam and Eve on our Earth, what about the millions of other Earths and their Adams and Eves? In a vision, Enoch spoke to the Lord, "yea, [there are] millions of earths like this, [and they] would not be a beginning to the number of thy creations."***[311] Did God wait for evolution to create bodies on millions of Earths? And did they all end up looking like God? The very thought is absurd in the extreme.

Returning to the Monte L. Bean evolution exhibit, "both Bybee and Jensen say they have had spiritual experiences as they've sought to understand and explain evolution."[312] Why did they say this? Was this to convince us of the correctness of their views? I, too, have had spiritual experiences as I have sought to understand evolution, and my views are staunchly anti-evolutionary. I ultimately trust that President Joseph F. Smith and the First Presidency had their own spiritual experiences as they prepared *The Origin of Man* as did President Heber J. Grant and the First Presidency as they prepared *The Mormon View of Evolution*. See Appendix 1.

---

[311] Moses 7:30

[312] Kristen L. Evans, *Evolution on Display at BYU*, BYU Magazine, Summer 2019, p. 12

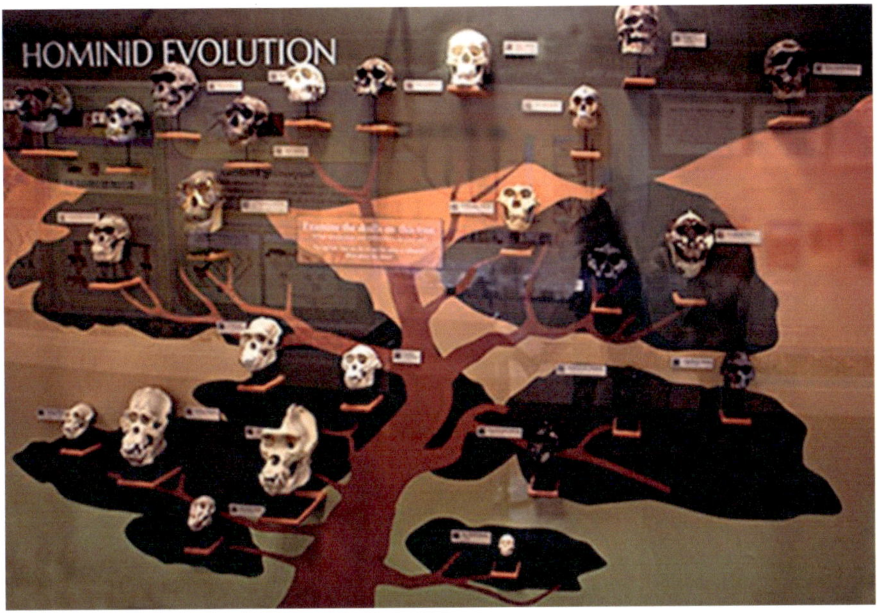

FIGURE 19. PROPOSED HUMAN EVOLUTIONARY TREE AT THE MONTE L. BEAN MUSEUM. [313]

As one leaves the museum, the following scripture is prominently displayed:

"And now, my sons, I speak unto you these things for your profit and learning; for there is a God, and he hath created all things, both the heavens and the earth, and all things that in them are, both things to act and things to be acted upon."[314]

More fitting scriptures might be the following.

"Behold, verily I say unto you, that there are many spirits which are false spirits, which have gone forth in the earth, deceiving the world."[315]

and

---

[313] Photograph by D. John Redman

[314] 2 Nephi 2:14

[315] D&C 50:2-3; see also D&C 45:57

"When they are learned they think they are wise, and they hearken not unto the counsel of God, for they set it aside, supposing they know of themselves, wherefore, their wisdom is foolishness and it profiteth them not." (2 Ne 9:28)

GLOSSARY - STRUCTURED

CELLS

Types:

Prokaryotes: these cells do not have an encapsulated nucleus; they make up two of the three domain classifications (Bacteria and Archaea)

Single-celled (protozoa): prokaryotes are all unicellular organisms

Eukaryotes: these cells have an encapsulated nucleus; they make up one of the three domain classifications (Eukarya); these organisms represent a small fraction of the living organisms on earth, however, due to their larger average mass their total biomass may be about the same as the total prokaryote mass

Single-celled (protozoa): there are single-celled organisms among eukaryotes as well as prokaryotes though these cells are more complicated; they have a nucleus and usually more organelles than prokaryotes

Multi-celled(metazoan): these multi-celled organisms make up all higher life forms, plants and animals being the most familiar; because there are multiple cells in these organisms, the cells can specialize into a vast array of cell types

Structure:

Cell wall: both prokaryotes and eukaryotes have cell walls, they provide structural support and protection; cell walls are very complicated, they are composed of lipids, carbohydrates, and proteins; cell walls need to control what goes into and out of the cell

Cytoplasm: everything inside of the cell wall and outside of the nucleus in eukaryotes

Life's chemicals: nucleic acids, proteins, lipids, carbohydrates, inorganic salts, and water (~80% of the cytoplasm is water)

Water: H_2O, water is ubiquitous in life, it has been called the universal solvent because it can dissolve so many substances (salts and hydrophilic organic molecules)

Polar: the water molecule has localized electrical charges, the electrons from the two hydrogens (H) atoms are attracted to the oxygen (O) atom, this makes the hydrogens slightly positive and the oxygen slightly negative

Hydrophobic: substances or objects that repel water or do not mix with water; water beads up on hydrophobic surfaces and does not wet the surface; these are generally non-polar substances

Hydrophilic: substances or objects that attract water, mix with water, or whose surfaces are wetted by water; these are generally polar substances

Organelles:

Nucleus: a walled organelle that contains most of the nucleic acid in eukaryotes

Ribosome: a walled organelle that manufactures proteins after mRNA enters their confines; it is a biological megamachine

Mitochondria: a membrane-bound organelle found in most eukaryotes whose primary function is to generate energy in the form of ATP (adenosine triphosphate) molecules for use in the cell; they also contain their own independent DNA which is inherited from one parent only – in humans, mitochondrial DNA usually is inherited from the mother; mitochondrial DNA is much smaller than nuclear DNA and, due to its one parent sourcing, makes it useful for scientific studies

LIFE'S CHEMICALS

Nucleic acids (DNA, RNA): these molecules contain all the information required for a cell to live, grow, and reproduce; they control most essential processes inside living cells

Proteins: these molecules are the workhorses of life; there are numerous types of proteins performing thousands of operations inside living cells

Lipids (fats): fatlike substances that are greasy, oily, and waxy and are not soluble in water; fats are one type of lipid

Carbohydrates (sugars): these are sugars, starches, and fibers; they are primarily used by cells as energy sources and for structural components

Polymer: a very large molecule comprised of repeating subunits

Monomer: a single subunit

Biopolymer: a polymer found in living organisms

Polynucleotide: a biopolymer chain of nucleotide monomers

Polypeptide: chains of amino acids linked by peptide bonds (> 50 amino acids is classified as a protein)

Chemical bonds

Ionic: a bond formed by an electrostatic attraction between oppositely charged atoms; electrons from the atoms or molecules are not shared between them

Covalent: a bond formed by sharing electrons between atoms, the positively charged nuclei of both atoms are both attracted to the same shared electrons

Hydrogen: an electrostatic attraction between a slightly positively charged Hydrogen atom in one molecule and a slightly negatively charged Nitrogen, Oxygen, or Florine atom in another molecule

Peptide: a C of one amino acid (from the carboxyl group, -COOH) covalently bonds to the N of another amino acid (from the amine group, $-NH_2$), releasing a water molecule (H_2O)

EVOLUTION

Study of

Taxonomy: the study of physical characteristics as a prelude to classification

Classification: a system of classification of life forms from the most specific to the most general (species, genus, family, order, class, phylum, kingdom, domain)

Phylogeny: the study of how physical characteristics compare between organisms

Phylogenetics: the study of how genomes compare between organisms

Principles

Genetic variation (mutations): random changes to the genetic code of an organism; there are many ways these changes can occur, however, they are relatively rare

Neutral: most genetic mutations have no discernable effect on an organism, nevertheless, the original genetic code usually becomes damaged

Harmful: many genetic mutations are harmful to an organism; mutations to regulatory and operational genes are particularly harmful

Beneficial: very few mutations are beneficial; if the effects of a mutation aid an organism in a very particular way to survive better in a particular environment, it is termed beneficial

Natural selection (survival of the fittest): beneficial mutations render an organism more fit for a particular environment, hence, they tend to survive and reproduce better than the rest of their species; in many cases, the beneficial mutation permanently takes over the entire population and becomes the species' new standard genome

Inheritable: since mutations are genetic, they are generally passed on to an organism's descendants

Slow, gradual, continual: aka the uniformitarianism model

Punctuated equilibrium: one of many alternative evolutionary models created to account for the fact that genetic change is not usually slow, gradual, or continual; the idea is first stasis (no change or equilibrium), a violent event (punctuated environment), then a return to stasis

Common descent: each species is descended from its precursor species; ultimately, there was the singular, one-time, original life form from which every organism on earth descended

Types

Microevolution (adaptation): minor evolutionary changes, usually at the species or genus level; these adaptations of the more fit organisms are usually due to mutations in structural genes

Macroevolution (new body types): major evolutionary changes leading to organisms in new families or higher; these changes are due to new body plans, new phylogenic structures, etc.; macroevolution changes must be made to operational or regulatory genes because too many changes to structural genes would be required; changes to operational or regulatory genes, however, are usually catastrophic to the organism

Evolution (presumed increase in information): mutations to organisms have historically assumed that they increased the amount of useful genetic information; there are no documented examples of an increase in genetic information spontaneously occurring in nature

Devolution (loss of information): most survivable mutations are the results of mutations that destroy the original genetic information; if these mutations are beneficial in a particular environment, they are permanently frozen into the organism's new genome due to natural selection; generally, the original genetic information is not recoverable

DNA

Genome: an organism's complete genetic code (information) described as a string of base pairs

Chromosomes: a chromosome usually has multiple genes in fixed positions; multiple chromosomes are usually grouped for protection and to save cell space

Genes: a strand of DNA or RNA that encodes either an RNA molecule or protein; genes are composed of introns, exons, promoters, terminators, and other related regions; in general, one gene encodes one protein

Introns: non-coding part of the gene; introns help regulate the coding process; these are removed before coding

Exons: the part of the gene that contains the actual code

ESS: Exonic Splicing Enhancer, identifies the beginning of an exon

ESE: Exonic Splicing Silencer, identifies the ending of an exon

Types of Genes: there are three general types of genes

Regulatory: these genes turn operator genes on and off

Operational: these genes govern the transcription of structural genes

Structural: these genes build proteins (translation)

Structure

Polynucleotide: a string of nucleotides

Nucleotide: these are the basic building blocks of DNA and RNA, they are composed of three subunits: a sugar, a base, and a phosphate; the sugar and phosphate form the backbones of both DNA and RNA with the base sticking out; see Figure 20

Sugar: a 5-carbon based molecule; DNA uses deoxyribose ($C_5H_{10}O_4$, it has one less oxygen atom than ribose, $C_5H_{10}O_5$)

Base: a base incorporating nitrogen (DNA uses four bases: A = adenine, T = thymine, C = cytosine, G = guanine; A always pairs with T, C always pairs with G)

Phosphate: the molecule PO_4, composed of phosphorus and oxygen

Codons: a sequence of three bases that form a coding unit for creating proteins

Permutation: technically, a permutation of 4 things taken 3 at a time in which order matters; since there are 4 different bases, 3 bases form 64 unique codes ($4^3 = 4x4x4 = 64$); a codon selects from 20 standard amino acids used to make proteins

Double helix: DNA has two intertwined backbones which wind around each other creating a double helix structure

Coding half: One of the DNA's backbones or strands is kept sacred and is not tampered with, it contains the genetic code; when being copied; it constitutes the Leading Strand

Template half: this strand is the complement of the coding strand, meaning it has TAGC whereas the corresponding coding strand has ATCG; basically, this strand is coded backward versus the coding strand; it constitutes the Lagging Strand

Copying order: There is a string of five carbon atoms in the deoxyribose sugar backbone of the DNA (deoxyribose, $C_5H_{10}O_4$). If one circumnavigates the carbon string in a clockwise manner, 3' refers to the third carbon atom and 5' refers to the fifth carbon atom in the string. The DNA molecule always grows from the 5' end, that is, the next nucleotide's 3' site attaches to the last 5' site in the growing DNA/RNA strand(s); see Figure 21

Copying the Leading Strand: the leading strand is copied in direct order (from 5' to 3') in a continuous stream

Copying the Lagging Strand: the lagging strand is complementary to the leading strand, consequently, it must be copied in reverse order; this is done by passing a segment of the lagging strand through the replisome, copying it backward (in direct order), spinning it around, and attaching it to the end of the growing lagging strand

Replication: the process in which a duplicate is made of DNA; the process includes copying, proofreading, and error correction; the error rate is about 1 error per 10 B base pairs

Replisome: a biochemical megamachine that does the primary copying of DNA; the DNA must be locally unwound, copied, and rewound; it must copy both the leading and lagging strands; it is composed of hundreds of proteins, specialized RNA molecules, and other molecules

Self-assembly/self-disassembly: the components of the replisome and other such molecular complexes are free-floating in the cell's cytoplasm; they are continually bumping into other molecules in the cell; when they reach a target site, they attach (self-assemble), accomplish their task, then detach (self-disassemble), and return to the cytoplasm

Transcription: DNA is copied to various types of RNA, a minimum of one gene is copied

Translation: after transcription of DNA into RNA, if the gene is destined to create a protein, various cutting, editing, and splicing steps must take place (such as removing introns) to create mRNA (messenger RNA); the mRNA is then translated into a protein's amino acid chain, i.e., each codon is translated into one of the 20 standard amino acids

Spliceosome: a biochemical megamachine that sequentially cuts the transcribed pre-mRNA at an appropriate location, removes introns from the RNA, and splices the exons back together; the components of the spliceosome complex self-assembles onto and self-disassembles from the RNA

RNA (SEE DNA)

Structure: RNA is single-stranded versus double-stranded DNA, with two other differences

Nucleotide: two differences versus DNA

Sugar: the backbone sugar is ribose ($C_5H_{10}O_5$) instead of deoxyribose ($C_5H_{10}O_4$)

Base: three of the bases are the same, U = Uracil replaces T, so bases are AUCG

Phosphate: identical to DNA

Encoding: RNA is always copied from the template strand of DNA which is the complement of the DNA coding strand, consequently, the RNA molecule is the complement of the complement (the reverse of the reverse) and is, therefore, identical to the DNA coding strand except for the base Thymine being replaced by Uracil

RNA World Scenario: according to Origin of Life researchers, the RNA World Scenario is the best proposal for how the spontaneous generation of life happened; present life forms require DNA and proteins to stay alive, unfortunately, there is a chicken-or-the-egg timing problem; which came first, DNA or proteins? RNA is the only biomolecule besides DNA and proteins that can occasionally replicate itself (aka DNA) and can catalyze itself in a restricted way (aka protein enzymes); these RNA enzymes are called ribozymes

PROTEINS

Structure: macromolecules consisting of chains of amino acids that fold into intricate, native, 3D structures; generally, there must be more than 20-30 amino acids to be considered a protein

Primary structure: the initial, unfolded sequence or chain of amino acids that arise from the DNA to mRNA transcription and translation process; the amino acids are held together by peptide bonds

Amino acids: chemically, amino acids contain an amine (-NH2), a carboxyl (-COOH), and a side chain which is unique to each amino acid; see Figure 22

Standard (protein-building) amino acids: 20 amino acids make up proteins, these are selected by the original DNA's codons

Secondary structure: the amino acid chain begins to fold into its native, 3D structure almost immediately after its creation, this folding is caused by chemical bonding between the various amino acids; these are held together by hydrogen bonds

Tertiary Structure: this is the final, native, 3D-shaped protein; the 3D structure is critical to protein function; often other proteins attach to a protein which can change its shape, after detachment, the original protein returns to its original shape; these shape-changing characteristics are the primary means by which protein complexes move like machines

Chaperone: many proteins have very complicated 3D shapes, the crowded and polar environment of the cell can complicate the folding process; chaperones are proteins that aid other proteins to fold correctly; if the chaperone becomes closed during the folding process, it is called a chaperonin

Quaternary Structure: this is a 3D structure that is an aggregate of two or more proteins; examples include the replisome, the spliceosome, and hemoglobin

Types of proteins: there are 8 types of proteins; these are listed along with an example in parentheses; the types are: structural (collagen, elastin), enzymatic/catalytic (blood clotting); hormonal (insulin), defensive (antibodies), storage (mineral ions, casein), transport (hemoglobin), receptor (insulin), and contractile (muscles)

EARTH FORMATION

Uniformitarianism: also called gradualism; this idea was popularized by Charles Lyell in his 3-volume work, Principles of Geology (1830); Charles Darwin had Volume 1 with him on his journey on the HMS Beagle (1831-1836); it is the theory that present natural forces operate the same as they did in the past and that things happened slowly, gradually, and continually

Catastrophism: the opposite of uniformitarianism; it is the theory that geologic and evolutionary changes are caused by sudden, violent events

FORCES OF NATURE

Forces of nature: there are four forces of nature, each force has one or more force exchange particles or force carrier particles

Electromagnetic: a well-known force in which opposite electrical charges (+, -) or poles (N, S) attract and like charges or poles repel; moving charges interact with magnetic fields, and changing magnetic fields exert forces on electrical charges

Photon: the force exchange particle for the electromagnetic force; also known as a particle of light

Strong: the nucleons (particles that make up the nucleus of atoms, e.g., protons and neutrons) are composed of smaller particles called quarks; quarks are held together by the strong force; it is called the strong force because it is stronger than any of the other three forces; the strong force must be stronger to overcome electrical repulsion forces inside nucleons and nuclei

Gluon: the force exchange particle for the strong force

Weak: the least well-known force of nature, it is involved in the radioactive decay of atomic nuclei; it is weaker than either the strong force or the electromagnetic force

W & Z bosons: the force exchange particles for the weak force; bosons are subatomic particles than have unitary quantum spins (0, ±1) such as all of the force carrier particles; NOTE: fermions are subatomic particles that have half-integral quantum spins (±1/2) such as electrons

Gravity: the most well-known force of nature; it attracts masses to each other

Graviton: a hypothetical force carrier particle for gravity

Theories: electricity and magnetism have been united into the electromagnetic force; subsequently, the weak force was combined with the electromagnetic force to create the electroweak force; the hope is for a Grand Unified Theory (GUT) that combines the electroweak and the strong forces; a Theory of Everything (TOE) would combine all four forces into one force theory

THERMODYNAMICS

Theory of Thermodynamics:

Entropy: is a measure of disorder, randomness, or lack of information in a system; crystals (e.g., quartz and diamond) have extremely low entropy, gases (e.g., oxygen and nitrogen) have extremely high entropy

System type: a thermodynamic system is a collection of matter and energy confined by a boundary, either physical or mathematical; the state of the system is defined by thermodynamic variables such an entropy

Isolated: a system that is isolated from the introduction of matter and energy (heat or radiation)

Closed: a system that can receive energy (heat or radiation) but not mass; operationally, the earth is a closed system since it receives massive amounts of both heat and energy from the sun

Open: a system that can receive energy (heat or radiation) and mass; technically, the earth is an open system since it can receive mass as well as heat and radiation, however, the quantity of mass from cosmic dust, meteorites, etc. is relatively minuscule

2nd Law: the total entropy of interacting thermodynamic systems always increases (+ entropy); this implies that natural processes are irreversible as this would mean that the change in entropy would be negative; there is also a tendency towards spatial homogeneity of both energy and mass, i.e., matter and energy tend to spread out and become spatially uniform

Equilibrium: a thermodynamic system is in equilibrium when there is no flow of matter or energy or the flows are small enough and slow enough to be considered in equilibrium; the earth is not in thermodynamic equilibrium

INTELLIGENT DESIGN

Intelligent Design (ID): the theory that states that some or all the features of the universe and living things is best explained by an intelligent cause, not an undirected process like evolution (physical forces and random mutations plus natural selection); most scientists call intelligent design pseudoscience

Creationism: is a religious belief that God is involved in the nature of the universe, the earth, life, and humankind; the belief that there was a special creation of nature and life; this paradigm may incorporate theistic evolution (God used evolution to create life forms) to reconcile religion and science; most scientists call creationism pseudoscience

Machines/molecular machines: a mechanical or electrical assembly of parts that transmit forces, motion, and energy to something in a predetermined manner; molecular machines are those made of molecules

Biomachine: a machine made of biomolecules such as proteins or RNA

Megamachine: an extremely large and complicated biomachine, e.g., the replisome and spliceosome

Dalton: a unit of mass for characterizing large molecules, approximately a mass in between the mass of one neutron and the mass of one proton; defined as 1/12 of a C_{12} atom (6 protons, 6 neutrons, 6 electrons)

Irreducible complexity (Michael Behe): the concept of irreducible complexity means a system of complementary, interacting parts that work together to perform some function; each part is required for the system to keep functioning, i.e., the system is irreducible – it cannot be made less complex and keep functioning; the system also cannot evolve and keep functioning – that would require that every part would have to change synchronously and in such a way as to keep working

Functional coherence (Douglas Axe): this concept means that all low-level functions of a system are completely aligned with the top-level functions of the system; it is related to irreducible complexity, that is, a higher-level machine may be composed of lower-level machines, each of which must be irreducibly complex and intimately tied to the function of the higher-level machine

Clockworks: the machine mechanism that runs a clock (mechanical clocks have numerous springs, gears, and bearings); extended to mean the inner workings of something; many biomachines resemble complicated clockworks, biomolecules performing the functions of springs, gears, bearings, etc. both figuratively and in actuality

Watchmaker: one who makes clocks or creates clockworks; the idea is that The Great Watchmaker (God) created the universe, the earth, life, and all of life's intricate molecular clockworks

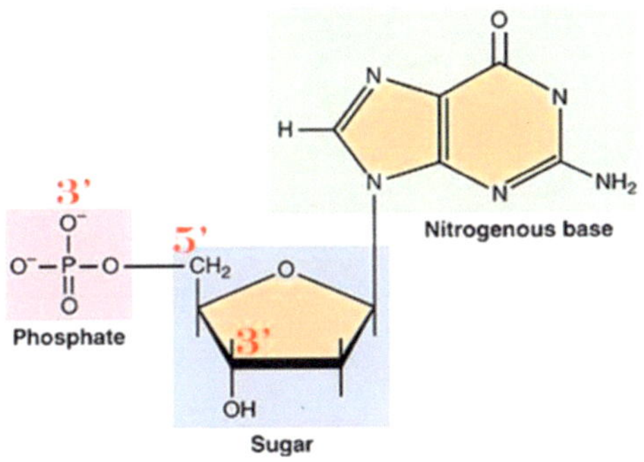

FIGURE 20. THE CHEMICAL DIAGRAM OF A NUCLEOTIDE. [316]

[316] Nucleotide, 0322 DNA Nucleotides.jpg, 18May2016; Anatomy and Physiology, OpenStax, May 18, 2016, CC BY 4.0; https://en.wikipedia.org/wiki/Nucleotide

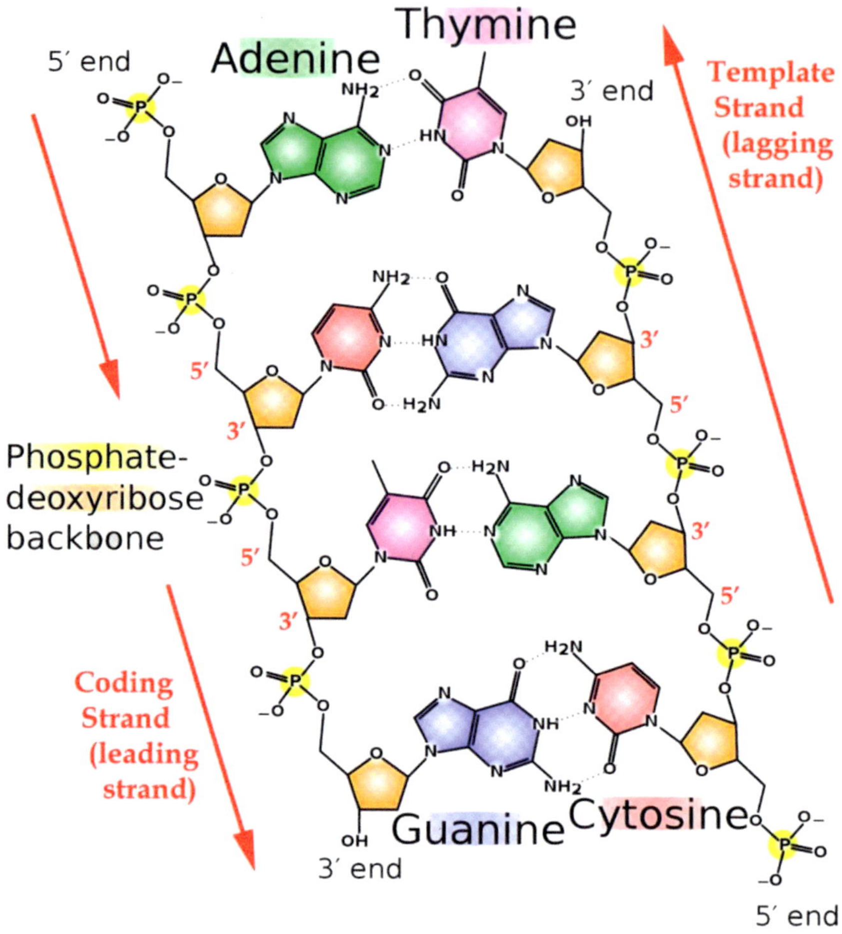

FIGURE 21. DNA/RNA BACKBONE DIRECTIONALITY (SYNTHESIZED IN THE 5' TO 3' DIRECTION). [317]

[317] Wikipedia: Directionality (molecular biology); DNA chemical structure.svg, Madeline Price Ball (Madprime), CC BY-SA 3.0, 20June2020; https://en.wikipedia.org/wiki/Directionality_%28molecular_biology%29

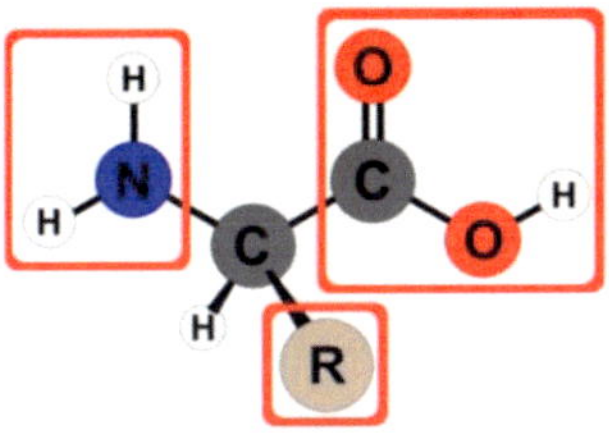

FIGURE 22. THE CHEMICAL DIAGRAM OF AN AMINO ACID (AMINE -NH2, CARBOXYL -COOH, AND SIDE CHAIN -R). [318]

[318] Amino acid, Amino Acid Structure.png, by Techguy78 - Own work, CC BY-SA 4.0, 17Apr2020; https://en.wikipedia.org/wiki/Amino_acid

REFERENCES

Alberts, B., Johnson, A., Lewis, J., et al, Molecular Biology of the Cell, 4th Edition, Garland Science (2002); extract found on the NCBI Bookshelf, a service of the National Library of Medicine, National Institutes of Health; https://www.ncbi.nlm.nih.gov/books/NBK26876/

All Things Bright and Beautiful, Children's Songbook, p. 231, The Church of Jesus Christ of Latter-day Saints, words and music by Cecil Frances Alexander, 1818-1895; Hymns for Little Children, 1848

American Scientist, Vol. 97, No. 3, May-Jun2009, *The Origin of Life*, Trefil, Morowitz and Smith, p. 206; https://www.americanscientist.org/article/the-origin-of-life

Andrus, Hyrum, God, Man, and the Universe, Bookcraft (1968)

Axe, David, *Estimating the Prevalence of Protein Sequences Adopting Functional Enzyme Folds*, Journal of Molecular Biology, Vol. 341, Issue 5, 27 Aug 2004, pp. 1295-1315;

Axe, David, Undeniable: How Biology Confirms Our Intuition That Life is Designed, Harper One (2016)

Barry, Drew, Unseeable Biology, The Walter and Eliza Hall Institute; https://www.ted.com/talks/drew_berry_animations_of_unseeable_biology

Basic Biology: DNA Replication; https://basicbiology.net/micro/genetics/dna-replication

Behe, Michael J., Darwin Devolves: The New Science About DNA That Challenges Evolution, Harper One (2019)

Behe, Michael J., Darwin's Black Box: The Biochemical Challenge to Evolution, Free Press (1996, 2008)

Behe, Michael J., The Edge of Evolution: The Search for the Limits of Darwinism, Free Press (2007)

Berlinski, David, The Deniable Darwin, Discovery Institute (2009)

Bostrum, Nick, Anthropic Bias: Observation Selection Effects in Science and Philosophy, Routledge (2010)

Bostrum, Nick, Philosophical Quarterly, 53, 2003, pp. 243-255

Broom, Neil, How Blind is the Watchmaker: Natures Design and the Limits of Naturalistic Science, Intervarsity (2001)

Brown-Driver-Briggs, Hebrew definitions; http://judaisztika.hu/downloads/BDB.pdf

Callaway, Ellen, *Scientists Synthesize Bacteria with Smallest Genome Yet*, Nature, March 25, 2016; cited in Scientific American, https://www.scientificamerican.com/article/scientists-synthesize-bacteria-with-smallest-genome-yet/

Canon, George Q., Journal of Discourses, 26 Vols. & Index, (abbreviated JD); 1:331; 6:31: 7:163; 7:285-6; 8:8; 9:317; 11:121; 15:137; 17:143; 21:201; Note: "The Journal of Discourses is not an official publication of The Church of Jesus Christ of Latter-day Saints, ... The Journal of Discourses includes interesting and insightful teachings by early church leaders (1854-1886); however, by itself, it is not an authoritative source of Church doctrine." https://www.churchofjesuschrist.org/study/manual/gospel-topics/journal-of-discourses?lang=eng

Chen, Weijun and Moore, Melissa J., *Spliceosomes*, Current Biology, Vol. 25, No. 5, pp. 181-3

Cold Springs Harbor Laboratory: DNA Learning Center; https://dnalc.cshl.edu/view/16938-3d-animation-of-rna-splicing.html

Continents Adrift, W. H. Freeman (1972)

Council of Europe; https://eclj.org/COE/council-of-europe-passes-resolution-to-ban-creationism-from-classroom

Coyne, Jerry A., see Wells, Jonathan, *"Shut Up," Coyne Explained*, Evolution News, January 19, 2012; https://evolutionnews.org/2012/01/shut_up_coyne_e/

Christensen, Matthew B., The First Vision: A Harmonization of 10 Accounts from the Sacred Grove, Cedar Fort (2014)

Darwin, On the Origin of the Species, 1859; http://darwin-online.org.uk/Variorum/1859/1859-188-dns.html

Dawkins, Richard The Blind Watchmaker: Why the Evidence of Evolution Reveals a Universe Without Design, WW Norton & Company (1996, 2015)

Dawkins, Richard, *Put Your Money on Evolution*. The New York Times Review of Books: 1989-04-09, p. 35

Dawkins, Richard, Speech at the Edinburgh International Science Festival (April 15, 1992); quoted in "*EDITORIAL: A scientist's case against God*", The Independent"(London), (p. 17), April 20, 1992.

Dawkins, Richard, The God Delusion, Mariner Books (2008)

Dawkins, Richard, *Time to Stand Up*, Freedom from Religion Foundation, September 22, 2001; http://ffrf.org/news/timely-topics/item/14035-time-to-stand-up

Delsemme, Armand H., Our Cosmic Origins: From the Big Bang to the Emergence of Life and Intelligence, Cambridge University Press (2001)

Dennett, Daniel, The Four Horseman: The Conversation that Sparked the Atheistic Revolution, Random House (2019); Christopher Hitchens, Richard Dawkins, Sam Harris, and Daniel Dennett filmed a landmark discussion about modern atheism in 2007, this is the conversation in print

Dennett, Show Me The Science, The New York Times, August 28, 2005; https://www.nytimes.com/2005/08/28/opinion/show-me-the-science.html

Denton, Michael, Evolution: A Theory in Crisis – New Developments in Science Are Challenging Orthodox Darwinism, Adler & Adler (1985)

Denton, Michael, Evolution: Still A Theory in Crisis, Discovery Institute (2016)

Discovery Institute; https://www.discovery.org/id/

Encyclopedia of Evolutionary Biology, *Novel Structures in Animals*, 2016; https://www.sciencedirect.com/science/article/pii/B9780128000496003140

Encyclopedia of Mormonism (5 Vol. set), Macmillan Library Reference, 4th Printing (1992), Appendix 4, *The Origin of Man*, 1908 & 1925; online http://eom.byp.edu

Evans, Kristen L., *Evolution on Display at BYU*, BYU Magazine, Summer 2019, p. 12

Everts, Sarah, *Protein Folding: Much More Intricate Than We Thought,* ACS ES&T Engineering, Vol. 95, Issue 31, pp. 32-38; https://cen.acs.org/content/cen/articles/95/i31/Protein-folding-Much-intricate-thought.html

Evolution, a 7-hour PBS television documentary, September 24-27, 2001

Futuyma, Douglas Joel, *Evolution as Fact and Theory*, BIOS, Vol. 56, No. 1 (Mar. 1985), pp. 3-13; Meyer, Darwin's Doubt, p. ix quoting Futuyma p. 8

Futuyma, Douglas Joel, Evolutionary Biology, 4th Edition (2007)

Gelernter, David, *Giving Up Darwin – A fond farewell to a brilliant and a beautiful theory*, Claremont Review of Books, Vol. XIX, No. 2, Spring 2019, pp. 104-109

Genome Research 2019, Jan., 29(1), 64-73; https://www.ncbi.nlm.nih.gov/pmc/articles/PMC6314161/

Godbey, W. T., *Proteins*, An Introduction to Biotechnology (2014); extract in ScienceDirect, *Protein Quaternary Structure*; https://www.sciencedirect.com/topics/neuroscience/protein-quaternary-structure

Goodsell, David S., The Machinery of Life, 2nd Ed., Springer (2009)

Gould, Stephen Jay, *Is A New and General Theory of Evolution Emerging* (1980), Self-Organizing Systems: The Emergence of Order, Springer, pp. 113-130 (1988); Meyer, Darwin's Doubt, 2014, p. ix quoting Gould p. 120

Gould, Stephen Jay, The Panda's Thumb: More Reflections in Natural History, WW Norton & Co. (1980)

Hawking, Stephen, A Brief History of Time, Bantam Books (1988), pp. 7, 125

Hawking, Stephen, and Mlodinow, Leonard, The Grand Design, Bantam (2012)

Helmenstine, Anne Marie, *10 Interesting DNA Facts*, ThoughtCo.; https://www.thoughtco.com/interesting-dna-facts-608188

Hill, Andrew, *The Myths of Human Evolution*, a review of The Myths of Evolution by Niles Eldredge and Ian Tattersall, American Scientist, 72 (March-April 1984), p. 189

Hill, *The Gift of Taungs*, review of Hominid Evolution, edited by D. V. Tobias, Nature, 323, (18 September 1986), P. 209

Hitchens, Christopher, Hitch-22, Atlantic Books (2010)

Hood, *Marlowe, Sweeping gene survey reveals new facets of evolution*, PhysOrg, May 28, 2018.

Institute for Creation Research, *Thermodynamics and the Origin of Life – Part II*, April 1, 1978, Duane Gish

Johnson, Phillip E., Darwin On Trial: 20th Anniversary Edition, IVP Books (1993, 2010)

Journal of Cell Science, 2017, 130, pp. 2941-2

Klein, Martin J., *Thermodynamics in Einstein's Thought*, Science, Vol 157, Issue 3788, 04 August 1967, p. 509

Klinghoffer, David, Debating Darwin's Doubt: A Scientific Controversy that Can No Longer be Denied, Discovery Institute (2015)

Kuhn, Joseph, *Dissecting Darwin*, Baylor University Medical Center Proceedings. 25 (1), 2012, 41-47

Kuhn, Thomas, The Structure of Scientific Revolution: 50th Anniversary Edition, 4th Ed., University of Chicago Press (2012)

Larson, Stan, *The King Follett Discourse: A Newly Amalgamated Text*, BYU Studies 18, No. 2 (1978)

Lehigh University, Department of Biological Sciences; https://www.lehigh.edu/~inbios/News/evolution.html

Lennox, John C., God and Stephen Hawking: Whose Design is it Anyway?, Lion Hudson (2011)

Lubenow, Marvin L., Bones of Contention: A Creationist Assessment of Human Fossils, Baker Books (2004)

Luskin, Casey, *Specialized Retinal Cells Are a "Design Feature," Showing that the Argument for Suboptimal Design of the Eye "Is Folly,"* Evolution News & Science Today reporting on Phys.org article, Casey Luskin, August 8, 2014; https://evolutionnews.org/2014/08/physorg_special/; original paper: Amichai M. Labin, Shadi K. Safuri, Erez N. Ribak, and Ido Perlman, "Müller cells separate between wavelengths to improve day vision with minimal effect upon night vision," Nature Communications, DOI: 10.1038/ncomms5319 (July 8, 2014)

Mecham, Amelia Mecham, *The Monte L. Bean Museum: Understanding the Principles of Evolution*, BYU College of Life Sciences, Summer/Fall 2019, p. 14

Medicines for Malaria Venture; https://www.mmv.org/malaria-medicines/history-antimalarials

Meyer, Stephen C., Darwin's Doubt: The Explosive Origin of Animal Life and the Case for Intelligent Design (2014)

Meyer, Stephen C., Signature in the Cell – DNA and the Evidence for Intelligent Design, Harper One (2009)

Monte L. Bean Museum, Brigham Young University; https://mlbean.byu.edu/exhibits/understanding-the-principles-of-evolution

National Center for Science Education: Embryo illustrations; https://ncse.ngo/accuracy-embryo-illustrations

National Center for Science Education: Intelligent Design; https://ncse.ngo/what-intelligent-design-creationism

National Center for Science Education: website; https://ncse.ngo/

Nelson, Russell M., Speaking Today: *The Magnificence of Man*, Ensign, Jan. 1988

Newton, Isaac, Theological Notebook, Part 2

Newton, Opticks: Or a Treatise of the Reflections, Refractions, Inflections & Colours of Light, based on the 4th Ed., London, 1730; Dover Books (2012)

Baugh, Alexander L., Opening the Heavens: Accounts of Divine Manifestations, 1820-1844, BYU Press (2005), *Parting the Veil: Joseph Smith's Seventy-Six Documented Visionary Experiences*, pp. 265-326

Oyster, Clyde W., The Human Eye: Structure and Function, Oxford University Press (1999 revised)

Packer, Boyd K., *We Believe All That God Has Revealed*, Ensign, April 1974

Paley, William, Natural Theology, Oxford World's Classics (1802, 2008)

Palmer, Steven E., Vision Science – Photons to Phenomenology, The MIT Press (1999)

Pearcey, Nancy R. and Thaxton, Charles B., The Soul of Science: Christian Faith and Natural Philosophy, Vol. 16, Crossway Books (1994)

Pew Research Center: In Focus - Mormonism in Modern America; https://www.pewforum.org/2007/05/16/in-focus-mormonism-in-modern-america/

PHYS.ORG: Sweeping gene survey reals new facets of evolution; https://phys.org/news/2018-05-gene-survey-reveals-facets-evolution.html

PHYS.ORG; https://phys.org/

PhysOrg, 3 July 2016, *Researchers report on the complex machinery of the spliceosome*; https://phys.org/news/2018-07-complex-machinery-spliceosome.html

Reichenbach, Andreas & Franze, Kristian & Agte, Silke & Junek, Stephan & Grosche, Antje & Grosche, Jens & Savvinov, Alexey & Guck, Jochen & Skatchkov, Serguei, (2012), *Live Cells as Optical Fibers in the Vertebrate Retina*, Selected Topics on Optical Fiber Technology, DOI 10.5772/26945

Reybrouck, David Van, *Imaging and imagining the Neanderthal: the role of technical drawings in archaeology*, Antiquity, Vol. 72, No. 275, 1998, p. 63

Rhodes, Michael J. and Moody, J. Ward, *Astronomy and the Creation in the Book of Abraham*, Astronomy, Papyrus, and Covenant, Studies in the Book of Abraham, Ed. John Gee and Brian M. Hauglid, Foundation for Ancient Research and Mormon Studies (FARMS), Brigham Young University, Provo, Utah (2005). Note: FARMS is now the Neal A. Maxwell Institute for Religious Scholarship.

Sagan, Carl, The Demon-haunted World: Science As a Candle in the Dark, Random House (1996)

Schwartz, Steven, Visual Perception: A Clinical Orientation, 5th Ed., McGraw Hill Education (2017)

Scientific American; https://www.scientificamerican.com/article/scientists-synthesize-bacteria-with-smallest-genome-yet/

Seattle PI: Which Mechanism Ensures the Accuracy of DNA Replication? https://education.seattlepi.com/mechanism-ensures-accuracy-dna-replication-4818.html

Shakespeare, William, Hamlet (1.4), Marcellus to Horatio

Shermer, Michael, How We Believe: Science, Skepticism, and the Search for God, Macmillan (2003)

Shermer, Michael, Why Darwin Matters: The Case Against Intelligent Design, Macmillan (2007)

Simmons, Tina, *Isaac Newton in Quarantine*, American Thinker, May 2, 2020; https://www.americanthinker.com/articles/2020/05/isaac_newton_in_quarantine.html

Smith, Joseph F., Deseret Evening News, December 27, 1913, Sec. III, p. 7

Smith, Joseph Fielding and McConkie, Bruce R., Doctrines of Salvation: Sermons and Writings of Joseph Fielding Smith, 3 Vols., Deseret Book (1905, 1999)

Smith, Joseph Fielding, compiler, Teachings of the Prophet Joseph Smith, Deseret News Press (1965)

Smith, Joseph, History of The Church of Jesus Christ of Latter-day Saints, 2nd Edition, 8 Volumes, Deseret Book (1978)

Smith, Warren, Modern Optical Engineering, 4th Ed., McGraw Hill Education (2000, 2007)

Smithsonian Environmental Research Center: Change in the air; https://forces.si.edu/atmosphere/02_02_00.html

Snow, LeRoi C., Improvement Era, June 1919, p. 656

Spetner, Not By Chance! – Shattering the Modern Theory of Evolution, Judaica Press (1998, 2006)

Stoeckle, M. Y., and Thaler, D. S., *Why Should Mitochondria Define Species?*, Human Evolution, Vol. 33, No. 1-2 (2018)

Stratton, Clifford James, The Facsimiles in the Pearl of Great Price: The Prophet Joseph Smith's 'Explanations, CreateSpace Independent Publishing Platform (2011)

Taylor, *John, Origin and Destiny of Woman*, The Mormon, New York, August 29, 1857

Seuss, Dr., The Cat in the Hat, Random House Books for Young Readers (1957)

John Taylor, *Origin and Destiny of Woman*, The Mormon, New York, August 29, 1857

Thaxton, Bradley and Olsen, The Mystery of Life's Origin: Reassessing Current Theories, Allied Books Limited (1984); reprinted in Thaxton, Charles B., Bradley, Walter L., Olsen, Roger L. Tour, James, Meyer, Stephen, Wells, Jonathan, Gonzalez, Guillermo, Miller, Brian, and Klinghoffer, David, Mystery of Life's Origin – The Continuing Controversy, (2020)

The Book of Mormon: Another Testament of Jesus Christ, The Church of Jesus Christ of Latter-day Saints (2013)

The Church of Jesus Christ of Latter-day Saints, *The Family – A Proclamation to the World*, The First Presidency and the Council of the Twelve Apostles of the Church of Jesus Christ of Latter-day Saints, September 23, 1995, Salt Lake City, Utah; https://www.churchofjesuschrist.org/study/manual/the-family-a-proclamation-to-the-world/the-family-a-proclamation-to-the-world?lang=eng

The Church of Jesus Christ.org website, Gospel Topics: Plan of Salvation; https://www.churchofjesuschrist.org/study/manual/gospel-topics/plan-of-salvation?lang=eng

The Doctrine and Covenants, The Church of Jesus Christ of Latter-day Saints (2013)

The Holy Bible, King James Version (LDS Edition), The Church of Jesus Christ of Latter-day Saints (2013); NOTE: extracts from the Joseph Smith Translation (JST), also known as The Inspired Version of the Bible, are found in the LDS edition of the Bible as footnotes

The Pearl of Great Price, Church of Jesus Christ of Latter-day Saints (2013); contains the Book of Moses, Book of Abraham and Joseph Smith – History

The Walter and Eliza Hall Institute: Drew Barry; https://www.wehi.edu.au/people/drew-berry

Turner, Rodney, The Footstool of God: The Earth in Scripture and Prophecy, Grandin Publishing Co. (1995)

Wandell, Brian A., Foundations of Vision, Sinauer Associates, Inc. (1995)

Way, Michael, *What I cannot create, I do not understand*, Journal of Cell Science, 2017, 130, pp. 2941-2

Wells, Jonathan, *"Shut Up," Coyne Explained*, Evolution News, January 19, 2012; https://evolutionnews.org/2012/01/shut_up_coyne_e/

Wells, Jonathan, Icons of Evolution: Science or Myth? Why Much of What We Teach About Evolution is Wrong, Regnery Publishing (2000)

White, Michael, Isaac Newton – The Last Sorcerer, Basic Books (1997)

Wiki: Archaeoraptor; https://en.wikipedia.org/wiki/Archaeoraptor

Wikipedia: 1911 Brigham Young University modernism controversy; https://en.wikipedia.org/wiki/1911_Brigham_Young_University_modernism_controversy

Wikipedia: Common Descent; https://en.wikipedia.org/wiki/Common_descent

Wikipedia: Dead Sea Scrolls; https://en.wikipedia.org/wiki/Dead_Sea_Scrolls

Wikipedia: David Gelernter; https://en.wikipedia.org/wiki/David_Gelernter)

Wikipedia: Directionality – Molecular Biology; https://en.wikipedia.org/wiki/Directionality_%28molecular_biology%29

Wikipedia: Extinction Event; https://en.wikipedia.org/wiki/Extinction_event

Wikipedia: GroEL: https://en.wikipedia.org/wiki/GroEL

Wikipedia: Human Genome; https://en.wikipedia.org/wiki/Human_genome

Wikipedia: Intelligent Design; https://en.wikipedia.org/wiki/Intelligent_design

Wikipedia: John Dalton; https://en.wikipedia.org/wiki/John_Dalton

Wikipedia: Julian Huxley; https://en.wikipedia.org/wiki/Julian_Huxley

Wikipedia: Lucy (Australopithecus); https://en.wikipedia.org/wiki/Lucy_%28Australopithecus%29

Wikipedia: Malaria; https://en.wikipedia.org/wiki/Malaria

Wikipedia: Michael Denton; https://en.wikipedia.org/wiki/Michael_Denton

Wikipedia: Mormon views on evolution; https://en.wikipedia.org/wiki/Mormon_views_on_evolution#:~:text=The%20LDS%20Church%20has%20no,evolution%20is%20opposed%20to%20scriptural

Wikipedia: Replisome; https://en.wikipedia.org/wiki/Replisome

Wikipedia: Richard Dawkins; (https://en.wikipedia.org/wiki/Richard_Dawkins

Wikipedia: Russell M. Nelson; https://en.wikipedia.org/wiki/Russell_M._Nelson

Wikipedia: Stringent response; https://en.wikipedia.org/wiki/Stringent_response

Wikipedia: Stephen J. Gould; https://en.wikipedia.org/wiki/Stephen_Jay_Gould

Wikipedia: Thermodynamics; https://en.wikipedia.org/wiki/Thermodynamics

Wikipedia: Watchmaker Analogy; https://en.wikipedia.org/wiki/Watchmaker_analogy

Williams, George C., Natural selection: Domains, levels, and challenges, Oxford University Press (1992)

Yan, Chuangye & Hang, Jing & Wan, Ruixue & Huang, Min & Wong, Catherine C. L. & Shi, Yigong, *Structure of a yeast spliceosome at 3.6-angstrom resolution*, Science, 11 Sep 2015, https://science.sciencemag.org/content/349/6253/1182?ve=

Zhang, Xiaofeng, & Chuangye Yan, Jing Hang, Lorenzo I. Finci, Jianlin Lei, Yigong Shi, *An Atomic Structure of the Human Spliceosome*, Cell, Vol. 169, Issue 5, 18 May 2017; https://www.cell.com/fulltext/S0092-8674(17)30487-7

REFERENCES

Zimmer, Carl, Evolution: The Triumph of an Idea, Harper Perennial (2001)

SCRIPTURAL REFERENCES

https://www.churchofjesuschrist.org/study/scriptures/study-helps?lang=eng

Old Testament

Gen. 1:14-19

Exodus 14:21-22

Joshua 10:11, 13-14

Psalms 139:14

New Testament

Luke 3:38

1 Cor. 15:22

2 Tim. 3:7

JST 2 Peter 3:8

JST Rev 12:4

Book of Mormon

2 Nephi 2:11, 14, 25

Alma 30:15, 18, 40, 44

Alma 34:9

Ether 3:4-16

Doctrine & Covenants

D&C 29:3, 31-31, 36

D&C 45:57

D&C 50:2-3

D&C 76:22-24

D&C 77:2, 6-7

D&C 88:20

D&C 93:24, 29-30

D&C 101:32-33

D&C 131:7-8

D&C 138:50

Pearl of Great Price: Book of Moses

Moses 1:8, 10, 31-35, 39-41

Moses 2:14-19, 26

Moses 3:7-8, 16-17

Moses 4:4

Moses 6:22, 59, 63

Moses 7:13-14, 30

Pearl of Great Price: Book of Abraham

Facsimile 2, Explanation, Fig. 1

Abr. 3:3-4, 9-10, 16

Abr. 4:14-19

Abr. 5:7-8, 13

Pearl of Great Price: Joseph Smith - History

JS-H 1:16-17, 30-32, 43

LIST OF RECOMMENDED READINGS

Issues with Evolution, Intelligent Design, the Earth

Axe, David, Undeniable: How Biology Confirms Our Intuition That Life is Designed, Harper One (2016)

Behe, Michael J., Darwin Devolves: The New Science About DNA That Challenges Evolution (Harper One, 2019),

Behe, Michael J., Darwin's Black Box: The Biochemical Challenge to Evolution (Simon & Schuster, 1996)

Behe, Michael J., The Edge of Evolution: The Search for the Limits of Darwinism (Free Press, 2007)

Clark, David P. & Russell, Lonnie D., Molecular Biology Made Simple and Fun, 4th Edition (Cache River Press, 2010)

Denton, Michael, Evolution: A Theory in Crisis (Adler & Adler Publ., 1986)

Denton, Michael, Evolution: Still a Theory in Crisis (Discovery Institute, 2016)

Gelernter, David, *Giving Up Darwin – A fond farewell to a brilliant and a beautiful theory*, Claremont Review of Books, Vol. XIX, No. 2, Spring 2019, pp. 104-109

Getting the Facts Straight – A Viewers Guide to PBS's *Evolution* (Discovery Institute Press, 2001)

Lo, Chien, Anderson, Alston, Waltzer, Evolution and Intelligent Design in a Nutshell, Discovery Institute Press (2020)

Lubenow, Marvin L., Bones of Contention: A Creationist Assessment of Human Fossils (Baker Books, 2004)

Meyer, Stephen C., Darwin's Doubt: The Explosive Origin of Animal Life and the Case for Intelligent Design (Harper One, 2014)

Meyer, Stephen C., Signature in the Cell: DNA and Evidence for Intelligent Design (Harper One, 2010)

Spetner, Lee M., Not By Chance!, Shattering the Modern Theory of Evolution (Judaica Press, 2006)

Thaxton, Bradley, Olsen, Tour, Meyer, Wells, Gonzalez, Miller, The Mystery of Life's Origin: The Continuing Controversy, Discovery Institute Press (Seattle, 2020); Note: this book includes an updated version of Charles Thaxton, Walter Bradley, Roger Olsen, The Mystery of Life's Origin, 1984)

Turner, Rodney, The Footstool of God: Earth in Scripture and Prophecy (Grandin Publishing Co., 1995)

Evolution, Scientific Naturalism

Dawkins, Richard, The Blind Watchmaker: Why the Evidence of Evolution Reveals a Universe Without Design (WW Norton & Co., 1996) – an example of evolutionary evangelism

Delsemme, Armand, Our Cosmic Origins: From the Big Bang to the Emergence of Life and Intelligence (Cambridge University Press, 1998) – a good overview of science's approach to our origins including cosmology

Goodsell, David S., The Machinery of Life, 2nd Edition (Copernicus Books, 2009) – relatively neutral, contains beautiful images

Zimmer, Carl, Evolution – The Triumph of an Idea (Harper Perennial, 2006) – a summary of evolutionary ideas

INDEX

ABOUT THE AUTHOR

Educational Experience

B.S. Physics, USAF Academy, 1970

M.S. Electrical Engineering, Arizona State University, 1975

M.S. Imaging & Color Science, Rochester Institute of Technology, 2008

Professional Experience

USAF, Senior Pilot, Instructor Pilot, Capt., 1970-1980

USAF Academy, Physics Instructor, 1978-1980

Hewlett-Packard Co., Engineer, Senior Engineer, Project Manager, Program Manager, Section Manager

Worked in HP Labs, R&D, Manufacturing, Materials Engineering, Quality Engineering, 1980-2012

Intel-Micron Flash Co., Engineer, Quality and Risk Assessment, 2014-2016

Church Experience

Gospel Doctrine, Elders Quorum and High Priests Quorum Instructor, 35 yrs.; High Priests Group Leader, Elders Quorum Presidency, YM President, Sunday School President, Stake Sunday School Presidency, Cub Master, Scout Master, Stake Missionary (3X), Temple Ordinance Worker

Family

Married to my soulmate, Pamela S. Stapley, for 49 years

7 choice children, 15 adorable grandchildren

Hobbies

Lapidary, Softball, Swimming, Hiking, Shooting, Gospel Study

IMAGE CREDITS FOR COVER PAGES

Earth photo: NASA, www.nasa.gov. Moon photo: the far side of the moon, photographed by Apollo 16, NASA; http://www.wikiwand.com/en/Far_side_of_the_Moon.

Nebula photo: Monkey Head Nebula, Hubble Space Telescope, 2011, NASA; https://hubblesite.org/image/3842/printshop.

The Creation of Adam, by Michelangelo, Sistine Chapel: Pixabay, free for commercial use – no attribution required; https://pixabay.com/photos/art-painting-mural-michelangelo-285919/.

DNA line drawing: DNA helix mode, Pixabay, free for commercial use – no attribution required; https://pixabay.com/illustrations/dna-molecules-biology-module-1015661/.

DNA image: space-filling model animation of D-DNA, made with qutemol, 14 May 2007, author Spiffistan, file Bdna.gif, Wikipedia commons;

https://commons.wikimedia.org/wiki/Main_Page.

Clockworks diagram: drawing of Harrison's H4 chronometer of 1761, published in The principles of Mr. Harrison's time-keeper, 1767, public domain;
https://en.wikipedia.org/wiki/Marine_chronometer.

"Jehovah Creates the Earth," is by Walter Rane © By Intellectual Reserve, Inc. This material is neither made, provided, approved, nor endorsed by Intellectual Reserve, Inc., or The Church of Jesus Christ of Latter-day Saints. Any content or opinions expressed, implied, or included in or with the material are solely those of the owner and not those of Intellectual Reserve, Inc. or The Church of Jesus Christ of Latter-day Saints; https://www.churchofjesuschrist.org/media/image/jehovah-creates-earth-rane-fa3141a?lang=eng

Made in the USA
Las Vegas, NV
03 November 2026

57422186R00150